Managing Risk:
The Health and Safety Contribution

Managing Risk:
The Health and Safety Contribution

Editor
John Stevens
Managing Director, RiskFrisk®

Tottel
publishing

Tottel Publishing, Maxwelton House, 41–43 Boltro Road, Haywards Heath, West Sussex RH16 1BJ

© Tottel Publishing 2005

A CIP Catalogue record for this book is available from the British Library.

ISBN 1 84592 048 1

Typeset by Wearset Ltd, Boldon, Tyne and Wear
Printed by Thomson Litho Ltd, East Kilbride, Scotland

Foreword

In a fast moving marketplace, risk management can sometimes become a reactive exercise or one of compliance rather than an effective management tool. But the climate for risk management is changing. Around the world, organisations are taking a fresh, hard look at risk and redefining the role of risk management in achieving performance objectives, and ultimately in driving their future success.

Organisations are constantly changing: products and services, markets and technologies continually evolve; organisational change, such as downsizing, new acquisitions or mergers, create new risks. Continual change can make long-established business risk management strategies ineffective; downsizing, new acquisitions or merging create new risks.

Successful organisations, whether in the private or public sectors, should not seek to eliminate risk rather they should actively seek to harness and manage it to their own advantage. Risk is no longer just a threat, when managed effectively it is a powerful asset that delivers competitive advantage, adds value and enables an organisation to achieve its aims.

Across any organisation, Occupational Safety and Health (OHS) professionals have an important contribution to make to the management of risk within their organisations and ultimately to the overall success of those organisations. This publication provides a welcome insight into the role OHS professionals can play in the development and management of an effective risk management strategy.

Terry M Neville
Pro Vice-Chancellor and Director of Finance
University of Hertfordshire

Preface

We aim to provide inspiration, guidance and practical advice to Occupational Safety and Health (OSH) professionals about the contribution they can make to the management of risk within their organisations. We show how changing to a risk-based approach enables OSH professionals to enhance their contribution and add value to the organisation, increase their level of influence and provide a valuable input to organisational development. We also demonstrate how they can use a risk-based approach to develop themselves, both professionally and personally.

We describe how OSH professionals can contribute towards risk management, and introduce our approach for identifying and managing those risks within an organisation that OSH professionals can influence, either directly or indirectly.

Core concepts, including organisational factors, are described and referred to throughout the book, and we discuss how key aspects of managing OSH risks are vital to the creation of a strategy that will enable OSH professionals to increase their influence on an organisation.

Collectively, we have significant experience of developing, implementing and managing OSH management systems at all levels – strategic, tactical and operational – and adopting approaches that are risk-based and not just legally-compliant risk averse based. By risk-based we mean balancing the maximisation of organisational opportunities and the minimisation of risks, not the too often risk averse legally compliant approach. We have used our extensive experience of applying the approaches we describe, which enables the reader to take advantage of our experience and knowledge to benefit them both professionally and personally, and to increase the success of their organisations.

We also describe how OSH professionals can create initiatives to increase their influence at higher levels of management decision-making, and therefore their added value, all within a business and risk management-focused approach. Like many other parts of an organisation, the OSH function too often undertakes its activities in accordance with its own self-image and based on 'external' perceptions and expectations of its contribution. This can cause many OSH professionals to contribute in a limited way, ignoring the potential contribution they can make to their organisation and to their professional and personal development. By changing to a risk-based approach, the OSH function will be able to enhance its contribution, be increasingly seen as organisationally relevant and make a significant contribution to organisational development and the achievement of the organisation's strategy and objectives.

Increasingly, OSH professionals have to provide advice and guidance that is competent and organisationally relevant. There have been a number of recent prosecutions of OSH professionals, raising the question within the profession about whether OSH practitioners, as individuals, are increasingly at risk and vulnerable resulting from their job role. In this book we use the term 'OSH professional' as we believe that it describes a level of expertise and added value that goes beyond the usual term 'competency', as professionalism is only partly about competency. OSH practitioners require professionalism in other areas, eg communicating with and influencing senior management and being able to talk a business language to demonstrate to the organisation how OSH risk management can add value.

We believe that there are three groups that can benefit from the approaches outlined in this book:

1 Organisations that are looking for a new approach to managing their OSH risks within the overall context of business risk management (BRM) and corporate governance (CG).
2 In-house OSH professionals who wish to understand the concept of risk management and the enhanced contribution that can be made to add value to the business.
3 Specialist external advisers on OSH/BRM who are looking for an enhanced approach to assist their clients to manage their OSH risks.

Any management system for managing OSH risks needs to be business and commercially focused and relevant to an organisation but, as a minimum, the substantial legal requirements must be taken into account. Consequently, the contents of the book contain a mixture of our extensive experience of implementing OSH risk management systems in a wide variety of organisations, plus a description of the UK legal background to the management of OSH risks.

In addition, we have used case studies, checklists, flowcharts and tables to demonstrate and highlight key learning points and support action plan development.

List of contributors

John Stevens – Managing Director – RiskFrisk®

John is Managing Director of RiskFrisk®, which specialises in the development and implementation of strategic health and safety risk management solutions with a strong focus on organisational development, people management and corporate governance. RiskFrisk® diagnostics are used to identify organisational, business, operational and people risk factors that are crucial to the successful integration of health and safety risk management.

Before setting up the company in 1998, John was International Safety Director for The BOC Group Health Care Division where he was responsible for strategic planning and implementation of health and safety risk management initiatives. Programmes ranged from developing and implementing organisation-wide management systems for general and specific risks through to improving communications methods and management training. After joining the BOC Group, John held increasingly senior positions in human resources, where he introduced innovative programmes for linking people management to the needs of the business.

Prior to joining the BOC Group in 1980, John spent 20 years in nationalised industries and local government and a public sector trades union. He gained extensive experience in all aspects of people management and organisational development, working in human resources, employee relations, national negotiations and training and development.

Lawrence Bamber – Principle Consultant – RiskFrisk®

Lawrence is a nationally and internationally recognised risk management and health and safety professional. His extensive experience covers business strategy, management systems and training. He is an acknowledged expert on UK regulations and, using his 'total' expertise, he provides RiskFrisk® with a key 'Technical Competency'.

Lawrence is a past President of IOSH, and remains an active member of IOSH Council and other committees, including conference planning. He often represents IOSH overseas.

He has written over 80 papers on risk management and health and safety, and is a contributing author to numerous OSH/risk management textbooks, such as Ridley's *Safety at Work* and Tolley's *Health & Safety Handbook*.

Lawrence has worked as an OSH/risk management consultant with companies such as Aon and Norwich Union, providing support for a wide variety of businesses.

He is a Fellow of IOSH, a Fellow of the Institute of Risk Management, and a Registered Safety Practitioner.

Elvis Cotena – Principle Organisational Psychologist – RiskFrisk®

Elvis is an Occupational Psychologist, Lecturer in Psychology and management trainer. His main areas of interest and expertise are organisational development, risk management, particularly in how it relates to organisational development and cross-cultural business communications. Elvis lived in Japan for two years whilst helping to set up Unilever's first plant there. His role was to mediate cultural differences in business practices, attitudes and communication styles.

Elvis has also lived and worked in Australia where he was senior lecturer in Business Communications at the University of Technology, Sydney, and designed/delivered a wide range of extramural management training programmes. He continues to enjoy teaching psychology whilst also working as an associate to a number of organisational development and training organisations, including RiskFrisk®.

Contents

Table of Statutes

Paragraph references printed in **bold** type indicate where the Act is set out in part or in full.

Table of Statutory Instruments

CHAPTER 1

Occupational Safety and Health (OSH) risk management

INTRODUCTION

1.1 What is 'Occupational Safety and Health (OSH) risk management' and why do we use the term? Our description of OSH risk management involves the integrated risk management of the following aspects:

- OSH organisational and strategic risks.
- OSH tactical and operational risks.
- OSH professional risks.
- OSH personal risks.

1.2 We use the term 'OSH risk management' because we believe that the risks to be managed or supported by the OSH function are much broader than the function's typical activities at a tactical and operational level. OSH functions can make an enhanced contribution to the organisation's overall management of risk if it looks both outward and inward from its current position and contribution.

1.3 OSH organisational and strategic risks – this is an approach to OSH management that treats the organisation's OSH strategy and tactics as a high-level strategic factor that is a positive and active asset to be developed, not a passive cost.

1.4 OSH tactical and operational risks – these are risks created by the typical activities of the OSH function and the operational management of the function and its direct support for the organisation.

1.5 OSH professional risks – these are risks created as a result of the perceptions (often historical, and often created in other organisations) formed by managers and employees, of the role and contribution that is currently being made, or can be made, by the OSH function to the organisation, and to managers and employees on a personal level.

1.6 OSH personal risks – these are risks created by a lack of competency, se aprofessionalism, role and personal development that restricts the individual OSH professional from making an effective contribution to the organisation.

PRIMARY ROLE

1.7   We believe that the primary role of the OSH function is to focus on the above aspects to support the management of risks and organisational development and success. But what is an organisation and what support can the OSH function provide? Our definition of an organisation is:

'An organisation is a network of relationships between people who come together for a common purpose.'

1.8   Using the key words from the above definition we can see that OSH professionals can support an organisation in the following ways:

1   Help to identify and maintain its common purpose, strategy and objectives.
2   Help to manage the relationships.
3   Help to maximise opportunities and minimise risks to people in the network and the organisation as a whole.
4   Help to ensure that people in the organisation have the competencies, knowledge and skills required to achieve the purpose, including to build effective relationships and successful networks.

1.9   We believe that OSH professionals can make an enhanced contribution to the success of organisations by ensuring that there is an effective and beneficial structured process in three areas:

1   The organisation – by advising senior management about the benefits of making business risk management (BRM) part of the normal management process and supporting the enhancement of general internal control systems.
2   The OSH function – by making the identification, assessment and management of opportunities and risks a cornerstone of its approach.
3   OSH professionals – by creating a structured process for the professional development of the organisation's OSH team, so that their individual and collective contribution is enhanced.

1.10   In our experience many organisations are too often restricted in their approach to innovation and creative approaches to management challenges. They often also regard the OSH function as a support function rather than an added value business partnership opportunity that can provide a valuable focus to assist the organisation to balance the management of opportunities and the minimisation of risk.

1.11   Restricting the contribution that can be made by the OSH function has important legal, moral, financial and business implications. OSH risks within organisations that are not effectively controlled can have a significant impact because of the increasing dependency of related business and operational functions. Some organisations have realised the criticality of key elements of their operation that are often not replicated elsewhere, either within the organisation or externally. As a consequence, some organisation have undertaken business continuity reviews – thinking through the 'physical' risks that face the business – but many have not considered the OSH risks that are often equally as crucial.

1.12   Every organisation's risk profile is different. The key issue for any organisation is to identify and measure its exposures and opportunities and manage them to the benefit of all stakeholders.

CORE CONCEPTS

1.13   There are five core concepts that are used as recurring themes:

1   Organisations should not restrict the management of risk to traditional areas, eg insurance or financial operations. OSH risk management strategies, and organisational policies and procedures, must be included in BRM systems. The organisation must ensure that those who are responsible for OSH risk management are actively involved in broader organisational BRM systems.

2   A prime responsibility of the OSH function should be to provide the effective management of risk for organisational activities that have an OSH dimension. OSH processes should support and be recognised and accepted as supporting the management of risk, and not dismissed as necessary legal compliance or 'red tape'.

3   OSH professionals should use BRM techniques of identifying risk exposures and evaluating control options to demonstrate how OSH risks can be managed and how their effective management adds value to the business. They must also demonstrate the cost of initiatives to improve the management of OSH risks compared to the cost of inaction.

4   OSH professionals must increasingly adopt a proactive risk-based and business focused approach rather than a reactive and risk averse legislative compliance approach. This will create an environment where they can develop their professional and personal competency and demonstrate added value.

5   OSH professionals must form business partnerships with other key influencers within the organisation eg financial management to support the minimisation of risk and the maximisation of opportunities. The proactive involvement of OSH professionals during an organisation's business strategy and planning processes is far more beneficial for the organisation, compared to a reactive involvement after the key decisions have been made.

IMPORTANCE OF OSH RISK MANAGEMENT

1.14   The effective application of legislation and the need for robust business relevant and integrated OSH risk management processes are crucial for all organisations. Regardless of what organisations say, in practice, OSH management is often not seen as core to the effective performance of the business. This can arise for three main reasons:

1   OSH is seen as merely a legal compliance function that makes sure that the organisation, as a minimum, is legally compliant.

2   OSH activities are often described, whether linked to legislation or not, as 'red tape'.

3   The reluctance of OSH professionals to get themselves on the 'risk and business agenda'. This may be either because of a concern that they may not be able to take a detached, independent view if they become too identified as part of the business and commercial aspects of the organisation, or because they do not have the necessary competency.

1.15   If OSH professionals can 'get on the business and risk agenda' within their organisations, then the risks of not managing the organisation's OSH risks come into sharper focus and the added value of the OSH function is much easier to demonstrate, be understood and welcomed.

1.16   The secret is for OSH professionals to think 'risk' when evaluating their contribution and to use a cost benefit approach that compares the cost of an intervention with the cost of inaction, ie leaving the risk uncontrolled. Using such an approach enables them to demonstrate how they can add value, rather than be perceived as a risk averse legislative focused function that only adds cost to the business. Changing the focus and perception of the OSH function can lead to new management approaches that identify different priorities and solutions, and thereby enhance the management of business and commercial opportunities.

ROLE OF OSH

1.17   Before we start to discuss BRM in general terms and how OSH professionals can become more involved, let us consider the role of the OSH function.

1   **Broader business agenda** – the management of risk is an essential part of corporate governance and corporate social responsibility (CSR), both concepts that are increasingly becoming the sign of a well-run and responsible organisation. The protection of all stakeholders' 'capital' is a vital aspect of good business management, and the key method is to use BRM techniques throughout the business process. OSH professionals should be ideally placed to influence appropriate management system developments, and create a proactive internal environment for risk identification and resolution. OSH professionals must seek to ensure that the organisation includes the concept of BRM within its normal management of the business. It is

vital that the identification and management of risk is not a reactive process when something goes wrong, but something that is considered in all aspects of an organisation. OSH professionals need to proactively contribute to the creation of an organisation that has the management of risk at the heart of its culture and business processes.

2   **Balancing opportunities and risks** – the organisation must ensure that its development of business opportunities within its strategic objectives must not be constrained by risk averse approaches, nor continued without due regard to the risks. The management of risk must be effectively balanced with the identification and management of opportunities. In addition, the culture of the organisation must not include human factors that cause people to adopt risk creating behaviours rather than intelligent risk-taking. OSH professionals must support the identification of negative hidden belief systems. For example, the organisation may say, 'Our people are our most important asset' – but in reality all the organisation's focus is on the achievement of financial targets. This leads to the hidden belief that the organisation's real value is 'Our people are an unimportant asset'. OSH professionals must assist the organisation to identify with the concept that risk exists in not recognising or not implementing an opportunity, rather than the traditional view that BRM is only about minimising losses before or after they occur.

One example where the OSH function can add value is in the discussion of outsourcing. When activities are being considered for outsourcing, eg car fleet management, site cleaning, planned preventive management (PPM), the OSH professional should challenge the organisation to think very carefully about the risk of outsourcing activities which have a crucial OSH element or can create risks that are under control if the activity is managed directly, but where control is reduced if the activity is outsourced. In our experience, as more and more activities are outsourced, the level of internal resources and expertise is reduced, creating an increasing reliance on the outsourced provider.

The assumed cost-benefit of outsourcing therefore may well be outweighed by a loss of control and a reduction in internal efficiency and risk management, and an inability to effectively manage the OSH risks, especially where there is a heavy people element that can be core to effective performance.

3   **Using key OSH skills** – OSH professionals must use skills of facilitation and relationship building to ensure that the organisation maintains an open culture and a challenging corporate 'mind' that it is prepared to operate as a matrix of opportunity 'teams', rather than each function within the organisation protecting its own responsibility areas and not thinking about or contributing to the overall strategic objectives.

OSH professionals should seek out and create business partnerships with those parts of the business that have influence through other mechanisms, eg financial control, human resources, legal and company secretariat. In that way they will increase their influence via direct and indirect routes.

4   **OSH role development** – OSH professionals should seek to use BRM to demonstrate added value, by identifying and managing risks across the whole range of OSH activities. OSH professionals need to get the basic legal compliance activities optimised. Once the most cost-effective and efficient systems are in place, they can concentrate on increasing their contribution to the organisation, by adopting proactive approaches rather than reactive responses to organisational developments. OSH professionals can contribute to a much greater degree if they use tools and techniques that demonstrate their BRM credentials and their business and commercial awareness and sensitivity. An increased contribution will enhance the professional and personal development of the OSH professionals.

5   **OSH risk management** – if the OSH function is to develop its role and increase its added value, then it must actively identify the risks that it is supporting or can influence elsewhere within the organisation, create mechanisms for control and ensure that business managers have the skills/knowledge to manage them on an ongoing basis. The identification of risk must be linked to a 'lost opportunity' approach of identifying the direct and indirect cost of inactivity, compared to the cost of the activity. Each OSH supported activity has a cost of action, but equally has a cost of inaction. For example, there is a cost in developing managers to undertake effective management of OSH processes, but equally there is a cost of not providing them with the skills, so that risks are not managed, losses are created and accidents/injuries occur. Unfortunately, such intangible costs do not appear on the balance sheet, and therefore are often ignored during decision making.

TYPES OF OSH ACTIVITIES

1.18   OSH role development can also be viewed under different types of OSH activities, namely OSH operations, business focused and business partnerships.

OSH operations

1.19   Essentially this includes OSH administration, policies and procedures, and OSH internal systems, to ensure full understanding by the OSH team, consistency of application, etc. The OSH function will have a general identification of what can go wrong and how to minimise these risks. OSH professionals must ensure that they deliver these base activities optimally with a high level of customer satisfaction, otherwise their efforts to get involved in 'higher' level activities will have no creditability. They can increase their added value by managing these activities in a proactive manner – providing management with core information, such as loss and accident figures that they can use to improve their costs and reduce risk. Too often, OSH information or OSH relevant information that is potentially useful to managers is held on a computer or hard-copy system, under the control of OSH and/or Human Resources, but not extracted, analysed and shared with business managers so that the OSH professionals can work with them to influence a change in practice.

Case study

1.20 One organisation did not realise the level of absence arising from work-induced stress and employee turnover. When the level of absence and turnover was identified and costed under the absence and recruitment processes, the true level of the risk was identified. Management were then much more easily able to see the benefits of implementing a change in policy and the introduction of new work arrangements and management approach.

Business-focused OSH

1.21 As OSH professionals improve and optimise their operational activities, the organisation can begin to appreciate that they can add value to the business, but often they are still not 'welcome' in the boardroom. Consequently, they are only asked for a reactive input to business decisions after they have been finalised. They are often fighting a rearguard action and can only add minimal value if they are reacting to decisions that may have been taken without regard to the OSH implications. They often have to resort to using changes in legislation as a way to introduce necessary changes in business and operational processes, rather than being able to put these on the business agenda in a proactive manner. The OSH professional needs to be proactive in influencing the organisation to put OSH issues on the business agenda. They can demonstrate how they can add value by keeping tuned in to business developments in the organisation and getting themselves invited to management meetings to present proposals on the OSH aspects of these developments.

Business partnerships

1.22 By taking the initiatives described in the preceding section, OSH professionals can get themselves involved in the discussion and decisions on future business strategies during the active development stage – rather than picking up the pieces once the strategies or implementation plans have been decided. Their overall aim needs to be to get themselves established as valued members of the organisations management team in recognition of their contribution to the organisation. One measure of success is whether managers seek them out for one-on-one discussions and advice at an early stage of their thinking about new activities.

BUSINESS RISK MANAGEMENT (BRM)

Overview

1.23   The need for organisations to identify, assess and manage risk has never been greater. The fast-changing nature of the business environment, the speed with which products and services can be brought to market, and the way that established organisations can soon lose what was hitherto a strong market position has meant that all organisations, large and small, along with public bodies, need to think about the way their organisations are run and the inherent risks across their total operations.

1.24   Risk is not something that comes into play when something goes wrong. Risk exists throughout the business cycle, from identifying a new product/service idea right through research, development, 'production', into distribution and after-sales support.

1.25   BRM is a process consisting of well-defined steps that, taken in sequence, support better business decision-making by contributing to a greater insight into risks and their potential consequences, both positive and negative.

1.26   The process is as much about identifying and maximising opportunities as it is about minimising unplanned losses. An essential element of the BRM process is to ensure that the identified risks and their control processes are closely monitored. By adopting effective BRM tools and techniques, organisations, and in particular OSH professionals, can help to improve the management of the business and business performance that are linked to the organisation's strategy, goals and objectives. Competitive advantage can be achieved by effective BRM, especially the pro-active management of OSH risks.

Benefits of BRM

1.27   The benefits of BRM are many, and a selection is listed below:

- More effective strategic planning.
- Increased knowledge and understanding of exposure to risk and the ability to recognise and evaluate new opportunities.
- Better utilisation of resources.

- Strengthened culture for continued improvement and collective BRM.
- Creating a best practice, quality-focused and risk-aware organisation.
- A systematic, well-informed and thorough method of decision-making.
- Improving cost control.
- Reducing unplanned losses.
- Willingness for external review.
- Enhancement of shareholder value by maximising opportunities and minimising losses.
- Competitive advantage.
- Motivated employees.

BRM vocabulary

1.28   OSH professionals should seek to understand their organisation's BRM vocabulary or develop their own for use within the OSH function. If no organisational vocabulary currently exists, then by developing and using their own vocabulary, OSH professionals will be able to demonstrate credibility.

1.29   Before embarking on the creation of a new or separate vocabulary, OSH professionals should liaise with BRM specialists within their organisation to agree a common set of definitions that they could use without confusing traditional risk-focused functions, such as insurance, that may already have their own terminology.

1.30   To assist this process, we have included definitions from two leading risk management standards to aid your understanding of the terminology we use in the book. The first Standard (United Kingdom) views risk management as the management of potential negative consequences, whilst the latter Standard (Australian and New Zealand) – the world's first published BRM standard, and revised in 2004 – views risk management as the maximisation of opportunities and the minimisation of risk. As described earlier, our view is that OSH professionals should approach and contribute towards an organisation's BRM system, based on the balanced principle of the maximisation of opportunities and the minimisation of risks.

United Kingdom Standard

1.31   The Institute of Risk Management (IRM), The Association of Insurance and Risk Managers (AIRMIC) and ALARM (National Forum for Risk Management in the Public Sector) jointly published a UK Risk Management Standard in October 2002 to 'set the scene for coherent thinking and application in the ever-widening field of risk management'. The Standard is aligned to ISO/IEC Guide 73 – Risk Management Vocabulary and, as an

Appendix, contains the ISO vocabulary. We have extracted some of the key definitions:

- **Risk** = 'Combination of the probability of an event and its consequence.'
- **Risk management** = 'Co-ordinated activities to direct and control an organisation with regard to risk.'
- **Risk-management system** = 'Set of elements of an organisation's management system concerned with managing risk.'
- **Risk treatment** = 'Process of selection and implementation of measures to modify risk.'
- **Risk control** = 'Actions implementing risk management decisions.'
- **Residual risk** = 'Risk remaining after risk treatment.'

1.32   The above Standard says that the term 'risk' is generally used only to refer to negative consequences. The focus of this book, however, is towards a broader more balanced approach that also includes the management of potential opportunities within the process of risk management, whilst minimising risk within those opportunities.

Australian and New Zealand Standard

1.33   A joint Australian and New Zealand Risk Management Standard (AS/NZS 4360: 2004) also includes the following definitions that do, in general, take a broader, more balanced view of risk management:

- **Risk** = 'The chance of something happening that will have an impact upon objectives. It is measured in terms of consequences and likelihood.'
- **Risk management** = 'The culture, processes and structures that are directed towards the effective management of potential opportunities and adverse effects.'
- **Risk-management process** = 'The systematic application of management policies, procedures and practices to the tasks of establishing the context, identifying, analysing, evaluating, treating, monitoring and communicating risk.'
- **Risk treatment** = 'Selection and implementation of appropriate options for dealing with risk.'
- **Risk control** = 'That part of risk management that involves the implementation of policies, standards, procedures and physical changes to eliminate or minimise adverse risk.'
- **Residual risk** = 'The remaining level of risk after risk-treatment measures have been taken.'

Business risk management (BRM), corporate governance (CG) and OSH risk management

1.34 Corporate governance and effective internal control are vital ingredients of organisational processes that ensure risks associated with organisational, people, business and operational activities are managed for the benefit of all stakeholders.

1.35 As a consequence of an increasing focus on risk, organisations are beginning to realise that virtually none of their activities is risk free. Whether considering their people, their customers, suppliers, operations or their finances, there will always be risks arising from possible developments that could prevent or impede an organisation from realising its strategic objectives.

1.36 In general, BRM has been a process that large organisations, or those with a heavy insurance requirement, have concerned themselves with. In addition, the majority of references to BRM relate to either insurance linked activity or financial BRM. However, due to some recent high-profile cases, it is likely that future legislation will increasingly require all organisations – large and small, private and public – to take on board BRM concepts and include it within their strategic, tactical and normal day-to-day operational activities, rather than be reactive when things go wrong. There is increasing pressure for BRM processes to be extended beyond insurance and financial-based risks.

1.37 There is also a much-increased global focus on corporate governance as a result of significant failures in high-profile companies in many parts of the world, eg Enron and WorldCom. Effective corporate governance should also include the organisation's management of its OSH risks.

1.38 The Organisation for Economic Cooperation and Development (OECD) has published the following definition, which is widely accepted:

'The corporate governance structure specifies the distribution of rights and responsibilities among different participants in an organisation, such as, the board, managers, shareholders and other stakeholders, and spells out the rules and procedures for making decisions on corporate affairs. By doing this, it also provides the structure through which the organisation's objectives are set and the means of attaining those objectives and monitoring performance.'

1.39 The work of the OECD, and other global and national bodies in the major economies, clearly shows that corporate governance will become increasingly important in national and the global economies and key to the operation of all organisations, private and public, large and small alike.

1.40 The UK focus on corporate governance took a major step forward in September 1999, when the 'Turnbull Report' was published by The Institute of Chartered Accountants in England and Wales. The Chairman of the Working Party was Nigel Turnbull, who was supported by representatives from commercial organisations and professionals service organisations. The Report was entitled 'Internal Control – Guidance for Directors on the Combined Code'. The Combined Code is linked to the 'Listing Rules' of the UK London Stock Exchange that apply to companies listed on the Exchange.

1.41 The Combined Code of the Committee on Corporate Governance (Code), and the Turnbull Report (Turnbull) include several key statements, which are very relevant to the contribution that OSH professionals can make to BRM:

- Code – 'The board should maintain a sound system of internal control to safeguard shareholders' investment and the company's assets.'
- Turnbull – 'The guidance is based on the adoption by a company's board of a risk-based approach to establishing a sound system of internal control and reviewing its effectiveness. This should be incorporated by the company, within its normal management and governance processes. It should not be treated as a separate exercise undertaken to meet regulatory requirements.'
- Turnbull – 'A company's objectives, its internal organisation and the environment in which it operates are continually evolving and, as a result, the risks it faces are continually changing. A sound system of internal control therefore depends on a thorough and regular evaluation of the nature and extent of the risks to which the company is exposed. Since profits are, in part, the reward for successful risk-taking in business, the purpose of internal control is to help manage and control risk appropriately rather than eliminate it.'
- Turnbull – 'All employees have some responsibility for internal control as part of their accountability for achieving objectives. They, collectively, should have the necessary knowledge, skills, information and authority to establish, operate and monitor the system of internal control.'
- Turnbull – 'Do the company's culture, code of conduct, OSH policies and performance reward systems support the business objectives and risk management and internal control system?'
- Turnbull – 'Do people in the company have the knowledge, skills and tools to support the achievement of the company's objectives and to manage effectively risks to their achievement?'

1.42 Two recent reports in the UK have taken the debate even further. Reports by Derek Higgs, 'Review of the Role of Non-executive Directors', and by Sir Robert Smith, 'Audit Committees – Combined Code of Guidance', were published in January 2003, and led to a revised Combined Code being published in July 2003. The revised Code came into effect in November 2003 for listed companies reporting years beginning on or after 1 November 2003.

1.43 In May 2004, the government published a consultation document on 'Draft regulations on the Operating and Financial Review and Directors Report'. This was in response to the recommendations of the independent Company Law Review and the Accounts Modernisation Directive from the European Union (EU).

1.44 Paragraph 2.5 in part states:

> 'Directors deciding in good faith what would be most likely to promote the success of the company, taking account of a wide range of factors, within and outside the company, which are relevant to achieving its objectives and to an assessment of its business. These factors may well include the company's impact on the environment and on the wider community, and its relationships with employees, customers and suppliers.'

1.45 We believe that the management of OSH risks should be a key component of this review and report; that OSH professionals should emphasise that the management of OSH risks are actively considered; and that comment on these risks should be included in the *Operating and Financial Review* (OFR).

1.46 When finalised, the requirements will apply to financial years starting on or after 1 January 2005 (since moved to 1 April 2005) for large and medium-sized companies. Companies will be studying the consultation document and the final regulations to determine what actions they need to take.

1.47 We believe that OSH professionals can assist and contribute to the thinking and the preparation of the organisation's OFR by posing the following questions:

- Does your organisation currently undertake such a review and publish such a report?
- Does it include aspects related to the management of OSH risks?
- What OSH risks are significant for your organisation?
- Has your organisation started to review its current methods of capturing the necessary data, including data on human capital risks, to ensure it is able to comply with the new reporting requirements?
- Have you considered the matters relating to the management of OSH risks that could be appropriate for inclusion in the review and report?
- Have you been actively involved in the review and the broader discussions about the content of the report?
- Have you sought out and created new internal 'partnerships' to ensure that your contribution is timely and effective?

A COMPETITIVE WORLD

1.48 Increasingly all organisations need to appreciate that, in this competitive world, they must make sure that their employees are managed effectively to maximise their contribution to the organisation. Just concentrating on growth and profits ignores the moral, legal and significant financial losses that can be caused by poor management of risk, especially OSH risks.

1.49   OSH policies and practices that are not relevant to the organisation's strategy and operations, restrict rather than encourage creativity, create other restrictions, or do not add value, can seriously undermine the chances of motivating employees and meeting business targets.

1.50   OSH risk management aspects of organisations must be recognised and included as a core element of the delivery of the strategic vision of the organisation. Effective 'full employment' in the UK, coupled with the increasing scarcity of qualified and competent people within the labour market, and the large number of unfilled vacancies, is placing huge burdens on organisations that have not considered these aspects, particularly at the operational 'deliver' point.

1.51   People, especially knowledge workers, who have skills and experience that are in demand, are less loyal to the organisation, and have been heavily influenced by employers who took the view that 'jobs for life' were no longer good policy. The demise of final salary pension schemes, due to organisations regarding them as being too expensive, and the view that 'employees over 50 years of age' are no longer adding value, has also contributed to a reduction in loyalty. This has resulted in the risks in being an employer changing and becoming more complex.

1.52   Organisations in a 'knowledge-based' economy will have to be much more proactive and innovative in their management of people. The risk of relying on old-style practices is too serious to contemplate. The effective management of OSH can be a competitive advantage and can be a definite motivator for employees. Employees or potential employees are starting to look at the broader picture when deciding on the value of an employment relationship with an organisation. Part of that relationship is a focus on the health and safety of employees.

1.53   As a consequence, in western economies the changing balance of power and changes in the relationships between the employer and the employed is much more fluid and dependent on economic conditions. This is often referred to as 'the changing psychological contract' between employers and the employed.

PSYCHOLOGICAL CONTRACT

1.54   The psychological contract, which has been widely researched and written about, influences employee's beliefs and behaviour in the workplace. From the recruitment stage of an employee's work life to retirement or resignation, it can have a profound effect on the attitudes and well-being of an individual. It is commonly understood as an individual's belief about the terms and conditions of a reciprocal exchange agreement with an employer or manager; a belief that some form of promise has been made and that the terms are accepted by all involved.

1.55 The psychological contract is an unwritten set of expectations between everyone in an organisation and, unlike the written contract, is continually changing. By nature it is a flexible and undefined set of terms, which may be interpreted by the individual in different ways at different stages in their working life. Although it is unwritten, it can be a significant determinant of behaviour in organisations, and perceptions of violation can have lasting effects.

1.56 When employees believe that their psychological contract has been broken, they often feel a great sense of injustice. Consequently, they are likely to reduce their contribution to their organisation both in terms of their own work performance and other 'good citizen' behaviours. These 'good citizen' behaviours can include compliance to and co-operation with OSH policies and procedures.

1.57 OSH professionals need to take the psychological contract into account when they advise their organisations about the impact on OSH risks of changes in organisational and operational processes. These concepts are particularly relevant when the organisation wishes to obtain the commitment of their employees to a change in OSH policies and practices.

ORGANISATIONAL FACTORS

1.58 Later in the book we will discuss how organisational factors can create risks, and how a review of organisational factors can be used for the identification and management of risk.

1.59 Organisational factors are organisational characteristics that can differentiate one organisation from another, and have a major impact on the behaviour and attitudes of employees. The creation of positive and proactive organisational factors is key to the success of an organisation, and must be managed as an active management process. Key factors are:

- **Vision statement** – a statement of the organisation's vision for its future (sometimes called a 'mission statement') that underpins everything it says and does.
- **Senior management commitment** – the extent to which management are committed to the management of risk and the management of an organisation's OSH risks.
- **Core values** – what particular aspects does the organisation focus on, what core values does it set for itself and its employees, and what hidden belief systems are actually in place, and influencing attitudes and behaviour?

- **Leadership style** – the energy and inspirational qualities of an organisation's leaders are a major factor in making an organisation one of the best to work for.
- **Organisational culture** – what value does the organisation place on the effective management of risk and, in particular, OSH risks?
- **Hidden belief systems** – what hidden belief systems are actually in place, and what influences attitudes and behaviour when no formal 'rules' are available?
- **Psychological contract** – to what extent is the 'unwritten' balance of the relationship between the organisation and the employee changing?
- **Responsibility framework** – to what extent are responsibility, authority and the allocation of resources delegated to the correct level at which action is most effective?
- **Occupational stress** – occupational stress has many negative consequences for organisations, and therefore it is a strategic risk that needs to be identified, quantified and managed.
- **Job/role design** – the ability of an organisation to create a structure, management style and work environment that encourages the best aspects of team working is a vital component in successfully managing any activity within the business, especially risk management.
- **Business strategy and goal setting process** – to what extent are these elements integrated and are goals and objectives formally set, cascaded, monitored and amended to take account of changing circumstances?
- **Performance management system** – to what extent is performance set, measured and managed to ensure active action-oriented activity?
- **Compensation and reward policies** – to what extent are compensation and reward policies, aligned to the business, linked to business goals and objectives, including risk management; and rewarding those behaviours that are consistent with the organisation core values, cultural framework and hidden belief systems?
- **Communication** – ensuring that communication systems are effective is a challenge in most organisations but the issue becomes ever more relevant when related to the management of risk. Effective communication is the foundation on which risk-management systems are based.
- **Learning and development** – the extent to which the organisational and individual needs for learning and development are considered as part of the means by which the organisation will achieve its objectives, and a key method used to enhance the contribution and productivity of the individual employee and teams.

DEVELOPING A RISK MANAGEMENT STRATEGY

Overview

1.60 As boards, directors and managers increasingly become accountable for losses and ineffective management of risks, they need a strategy that assures they understand the implications and risks associated with every decision they make. Without a BRM system in place, consequences and accountability can be a lottery!

1.61 Any organisation, private or public, large or small, can benefit from using risk management tools and techniques. Many large organisations have some form of BRM in place. However, very few involve OSH risks.

1.62 Every organisation's risk profile is different, even within the same sector. The crucial issue is to identify and measure the organisation's exposures and opportunities and manage them to the benefit of all stakeholders. In our experience the tools and techniques included within this book can be used by any organisation to this end.

1.63 The development of a process to identify OSH risks, and the creation of a strategy for their effective management, is covered in **Chapter 7**.

BRM PROCESSES AND OSH

1.64 Organisations simply cannot allocate responsibility for BRM to 'risk managers' and expect that the processes they are able to put in place will be welcomed, accepted and maintained by managers and employees who are already overburdened by their workload. Managers and employees are generally focused on day-to-day operations (driven by the organisation's focus on short-term 'production') and do not take a longer-term view or focus on identifying and managing intangible risks. Therefore, 'risk managers' cannot expect to get their concepts on the broader business agenda and incorporated into the way that the organisation runs if they take a risk averse approach. There is nothing more frustrating to managers who are heavily focused on business and commerce, than to have within their BRM functions or external advisors, people who do not understand the realities of business, and therefore advise on strategies and approaches that restrict effective risk managing decision-making.

1.65 As risk is inherent in any business decision/process, those within 'BRM functions' must advise on the potential risks, but must also seek to add value by offering alternatives that are consistent with the general strategic direction that the organisation has chosen to take. 'Risk' must be on the board agenda, but it must not become a 'bored' issue that serious business managers seek to avoid because unrealistic 'risk averse' approaches are all that is on offer. All managers within an organisation must accept the responsibility for managing risk within their area of responsibility, but equally must be given the skills and tools to do so. Additionally, the organisation must ensure that the full business process for each area of operations is considered when risk exposures are being identified.

1.66 Organisational factors identified in an earlier section generally have organisation-wide implications and therefore must be taken into account during the creation of BRM processes. The activities of the OSH function normally requires consideration of organisation-wide processes, but often the OSH function is not viewed as part of the general BRM system, and very often is not involved in BRM. In general, organisations do not see the OSH function as part of BRM, and OSH professionals would not typically see themselves as undertaking a BRM activity or contributing towards managing risk.

1.67 If the OSH function is to be included within BRM, then those whose primary function is to advise on BRM must be helped to understand and accept that the OSH function is part of BRM. The benefits of including the OSH function in BRM can include:

- Minimising OSH risks will contribute towards the success of the organisation, both in tangible ways, eg 'productivity', but also in intangible ways, eg morale and compliance with internal controls.
- OSH professionals can support the implementation of and compliance to internal controls especially BRM. The success of internal controls, including BRM, is ultimately dependent on people, not technology, both of which are key elements of the management of OSH risks.
- OSH professionals are ideally placed to assist with the implementation of internal controls, as they can take an organisation-wide perspective, and they can operate at all levels within an organisation – strategic, tactical and operation.
- OSH professionals can bring their expertise to bear in these areas to ensure that the BRM system goals and objectives, and OSH objectives, are integrated with performance-management and reward processes.
- OSH professionals should seek to use skills of influencing, facilitation, strategic thinking, non-silo-constrained thinking to assist in identifying innovative ways to ensure BRM systems are maintained.

1.68 The issue is that OSH functions do not typically see themselves as BRM functions but, in reality, that is one of their fundamental roles, and needs to be one of the core competences of OSH professionals.

1.69 The challenges facing OSH professionals are:

1   To influence the inclusion of BRM on the broader business agenda, particularly at board level, and influence the corporate governance/ internal control processes.

2 To assist management in developing the organisation to ensure that it effectively balances the management of its opportunities, whilst minimising its risks, thereby supporting the achievement of its strategic objectives.

3 To develop its role, and acceptance of the role change, to one of managing risk within the total range of its OSH activities, and to increase the added value of the OSH function across its range of activities.

4 To increase the influence on the business by changing the OSH role from 'legal compliance' to 'business partnership'.

TRUST AND BRM

1.70 Building successful working relationships and partnerships is of fundamental importance in managing risk in organisations. Trust is fundamental to creating these relationships. In the current climate of pressure for growth and profits, corporate greed, cost-cutting and changes in the psychological contract, which all have the potential to destroy trust, OSH professionals have a key role to play in acting as the agents of co-ordination of the tensions and conflicting interests in organisations. They are uniquely placed to encourage leaders and employees to invest sustained, focused attention on building and maintaining trust within the organisation.

1.71 Failure to build the levels of trust in organisations carries the risk that it could result in:

- Turf issues and less-than-effective working across organisational interfaces.
- Employee commitment and discretionary effort being suboptimal.
- Disconnection between the interests of the organisation and those of the employee.
- Loss of innovation, risk-taking and creative energy.
- Reduced organisational performance and increased operating costs.
- Reactive rather than proactive compliance to OSH policies.

1.72 Although these are significant risks that OSH professionals need to support the management against, there does not appear to be much evidence that OSH professionals recognise that building and maintaining trust needs to be a specific objective.

STRUCTURED AND INTEGRATED APPROACH

1.73 The effective management of OSH risks requires a structured approach to be adopted, and the resulting controls integrated with normal organisational and operational processes. Many organisations approach the management of risk either in a piecemeal manner, or reactively, following a loss.

1.74   In our experience, the most effective and well-managed organisations will use a structured approach that takes account of the profile of the organisation, will identify where they are exposed to risk, will evaluate existing control mechanisms, and will evaluate what more needs to be done to ensure that the risks are tolerable and effectively managed to the level decided.

1.75   The most effective BRM tools and techniques are those that are designed to provide answers to key questions at each stage of a structured process, and to enable the organisation to select the factors they wish to concentrate on during the next step.

1.76   In our experience, using a step-by-step approach enables the organisation to adapt the area of review and the depth of study to match the precise requirements of their organisation. In this way the necessary information is identified and decisions can be made about proportionate actions.

1.77   But often organisations do not have the resources to resolve all their risk exposures in one big exercise. Consequently, a step-by-step approach, that progressively concentrates the focus of the review, and at the same time increases the level of investigation, is an effective management approach.

1.78   A description of a structured process for building a 'OSH risk register' and a strategy for increasing the influence of OSH professionals on BRM is explained in **Chapter 7**.

SUMMARY OF REMAINING CHAPTERS

Chapter 2 – OSH and business risk management

1.79   This chapter explains that the management of risk is a vital and key part of managing any organisation. It briefly describes some important cases – *Herald of Free Enterprise*, Piper Alpha, etc – to show how each had a significant OSH element, that was not taken into account during more normal 'insurance/financial' focused business risk management processes to identify and control relevant risks.

1.80   We show how OSH must now be a key part of the corporate social responsibility agenda within organisations, which must relate to all stakeholders connected to the organisation's internal and external activities, including employees or others affected by the organisation's activities.

1.81   We show how effective corporate governance processes within an organisation must include OSH. How business decisions must consider all risks and consequences of a business strategy. We make reference to Turnbull, Higgs and recent changes in the Stock Exchange Combined Code in relation to annual reporting.

1.82   We explain that it is traditionally difficult for OSH professionals to be asked to participate in corporate social responsibility (CSR) and corporate governance (CG) processes, as they often only operate at the 'operational' level within organisations and, additionally, approach their roles in a risk-averse, non-business added value manner. We demonstrate that OSH professionals must seek to increase their influence 'up the management chain' so they are asked to contribute at the 'tactical' and 'strategy' levels, where business risk management is typically on the agenda. In this way, OSH professionals can help the organisation to manage its opportunities in a more complete manner, whilst minimising the risks.

Chapter 3 – OSH management systems

1.83 This chapter explains that the relationship to regulations – so favoured by many OSH professionals – does, in one key area, have a significant relevance to business risk management because of the existence of HSG 65, which describes how OSH should be built as a management system and fully integrated with normal business and operational processes of any organisation.

1.84 We show how OSH professionals must build a business case for action by using business risk management techniques to show the balance of cost verses risk. Organisations collect much data, which the OSH professional can use to show the current cost of inaction, compared to the cost of action. However, typically, OSH professionals just provide the cost of action relying on the 'strength/threat' of regulations to justify action. This is mostly a poor motivator for a positive business decision. OSH professionals must also resist the temptation to think that throwing money at an issue is the way to resolve the matter. Often the people or management approach is far more successful and creates a longer-term sustainable solution.

1.85 Our experience shows that employees are motivated to work well for an organisation that includes a strong focus on BRM, health, safety, human resources and training as part of the 'way they do business'.

1.86 As no workplace is exempt from a duty of care, it makes financial sense to protect employees and protect organisations from potential financial penalties, arising from organisational and operational risks, and accidents/ill-health at work. We pose the question: are your profits – and your reputation – sufficient to cover such losses?

1.87 As with the 'cost of quality', it is not always possible to accurately calculate the cost of losses and accidents. Consequently, managers tend to focus on reducing direct costs and minimising the costs of those measures needed to comply with regulations.

1.88 This leaves a whole range of costs (indirect costs) outside active consideration and therefore ignores the bulk of the costs and other resource implications. What is required is a complete understanding of the cost of losses. We pose another question: what level or type of resources do you need to control these losses?

Chapter 4 – OSH policy

1.89   We explain the key stages in the development of an effective policy:

P O P I M A R

P	olicy.
O	rganisation.
P	lanning.
I	mplementation.
M	onitoring.
A	udit.
R	eview.

1.90   It is vital that all stages are integrated and create a closed-loop process that provides feedback on system improvements to the policy-formation stage. If senior management are not provided with feedback then they will assume that the process, in its current form, is managing the risk effectively. Senior management like reassurance and do not like surprises!

Chapter 5 – OSH business processes

1.91   We explain that there are three key aspects to OSH processes for managing risk:

1   **Cover all business and operational processes** (internal and external), by ensuring that the OSH professional is asked for input to all aspects of the organisation, prior to final decisions being made. It is vital that facility-based and field-based activities are covered, plus work-related driving.

2   **Current** – identify the existing activities, and related hazards/risks, then effectively control the risks. The relationship to other related activities, eg property conservation, emergency planning and business continuity must be strong, especially if they are not managed by the OSH professional.

3   **Future** – manage changes to organisational and operational processes, especially where there are OSH implications, so that new/amended risks continue to be managed as part of the OSH system.

1.92   In addition, if the OSH function is to be taken seriously as a contributor to business risk management then it must approach its areas of responsibilities by identifying exposures and then demonstrating available controls and the benefits of each.

1.93 Typical OSH business processes are described, with examples of polices and processes for effective OSH, BRM and organisational integration.

Chapter 6 – Organisational and human factors

1.94 This chapter examines how effective BRM requires organisational and human factors to be managed as part of an overall organisational and operational systems approach.

1.95 'Organisational' factors are the structures and processes that largely determine the culture of an organisation and include factors such as responsibility frameworks, learning and development programmes and job/role design. 'Human' factors are fundamental human characteristics that need to be understood and managed as part of the BRM process. They include the nature of human error, how risk is perceived regardless of actual risk and what motivates individuals in the workplace to take risks.

1.96 The cross-functional nature of OSH and its focus on people and technology provide OSH professionals with the opportunity to become an integral and vital part of the BRM process.

Chapter 7 – Getting started – increasing your contribution

1.97 This chapter will show how OSH professionals can identify, assess and classify OSH risks. In addition, an explanation will be given of the creation of a strategy for the implementation of improvements in the management of OSH risks, thereby increasing the contribution of the OSH function and demonstrating its potential for enhanced added value.

1.98 The concept of a Risk Register will be described that can be used to list risks and potential risk treatments. Other key aspects covered include:

- OSH risk identification.
- Input, activity and output risk identification.
- Process approach to risk identification.
- Classification and risk assessment.
- Implementation challenges.
- Strategy for increasing contribution.

1.99 Particular reference will be made to the professional and personal development of OSH professionals.

Chapter 8 – Pulling it together

1.100 This chapter will pull together the key points from earlier chapters and summarises the way forward for OSH professionals to identify and assess OSH risks and enhance their contribution to BRM.

REFERENCES

The Accounts Modernisation Directive from the European Union (EU), Directive 2003/51/EC of the European Parliament and of the Council – 18 June 2003 amending Directives 78/660/EEC, 83/349/EEC, 86/635/EEC and 91/674/EEC on the annual and consolidated accounts of certain types of companies, banks and other financial institutions and insurance undertakings (http://www.europa.eu.int/eur-lex/pri/en/oj/dat/2003/1178/117820030717en 00160022. pdf)

Combined Code on Corporate Governance, UK Financial Services Authority, 2003. The Financial Reporting Councils' Combined Code – Listing Rules for UK Stock Exchange listed companies (http://www.frc.org.uk)

Draft regulations on the Operating and Financial Review and Directors' Report – A consultation document – UK Department of Trade and Industry, http://www.dti.gov.uk/cld/condocs.htm, May 2004

The Higgs Review, http://www.dti.gov.uk/cld/non_exec_review, 2003

The Organisation for Economic Cooperation and Development (OECD) OECED Principles of Corporate Governance, http://www.oecd.org/dataoeced, 2004

The Smith Report, http://www.frc.org.uk/publications/content/ACReport.pdf, 2003

Turnbull Report, Internal Control: Guidance for Directors on The Combined Code, The Institute of Chartered Accountants in England and Wales, http://www.icaew.co.uk/internalcontrol/, September 1999

White Paper, *Modernising Company Law*, http://www.dti.gov.uk/cld/modern/index/htm, published in July 2002 – Cm 5553 – I & II, July 200

Occupational Safety and Health, corporate governance and business risk management

INTRODUCTION

2.1 This chapter explains that the management of risk is a vital and key part of managing any organisation. It briefly describes some important cases – *Herald of Free Enterprise*, Piper Alpha, etc – to show how each had a significant Occupational Safety and Health (OSH) element that was not taken into account during more normal 'insurance/financial'-focused business risk management processes to identify and control business and operational risks. Additionally, if a business risk-management programme had been in place, and meaningful risk decisions had been taken further up the management tree with all relevant risk assessments in place, then rather than merely making decisions on an arbitrary cost basis, all factors could have been taken into account.

2.2 This so-called high-level 'cost only' decision-making is symptomatic of many board decisions taken without much thought for the risk side of the equation. Indeed, it shows a lack of understanding at senior management level of the meaning of the phrase 'so far as is reasonably practicable' – ie cost versus risk – which is enshrined in the Health and Safety at Work etc Act, 1974.

2.3 If nothing else, a business risk management (BRM) programme high-lights the importance of risk assessment to the board/Senior Management within organisations and ensures that both cost and risk are taken into account when management decisions are taken and implemented.

2.4 This chapter also looks at the chronological development of how the consideration of non-financial/operational risks at board level has been elevated in importance over the last few years, with Turnbull (see **Chapter 1**) being seen as something of a watershed.

2.5 We examine how a BRM programme helps to elevate the profile of OSH within an organisation's overall corporate governance (CG) and corporate social responsibility (CSR) approach and stresses the need for the 'safety net' to be extended to include all potential organisational stakeholders. It goes without saying that this extension will include ensuring that all stakeholders – eg contractors, suppliers – have similar OSH/BRM programmes operational within their organisations. If they have not, then, from a good neighbour viewpoint, coaching, mentoring and vetting may take place, especially if the particular stakeholder is one the organisation wishes to do business with! This is how competitive advantage can be created.

2.6 We examine the role that OSH professionals can play in BRM and also highlight how a proactive BRM programme can greatly assist organisations in achieving their CG and CSR goals and objectives. This is especially important as the Health and Safety Commission (HSC)/Health and Safety Executive (HSE) are taking an increasing interest in how Stock-Exchange-listed companies are reporting on CG/CSR/OSH issues in their Annual Reports! We will return to this subject in **Chapter 7**.

2.7 We show how OSH must now be a key part of the corporate social responsibility agenda within organisations, which must relate to all stake-holders related to the organisation's internal and external activities, including employees or others affected by the organisation's activities.

2.8 We show how effective CG processes within an organisation must include OSH. How business decisions must consider all risks and con-sequences of a business strategy. We make reference to Turnbull, Higgs and recent changes in the Stock Exchange Combined Code in relation to annual reporting.

2.9 We explain that it is traditionally difficult for OSH professionals to be asked to participate in CSR and CG processes, as they often only operate at the 'operational' level within organisations and, additionally, approach their roles in a risk-averse, non-business added value manner. We demonstrate that OSH professionals must seek to increase their influence 'up the management chain' so they are asked to contribute at the 'tactical' and 'strategy' levels, where business risk management is typically on the agenda. In this way, OSH professionals can help the organisation to manage its opportunities in a more complete manner, whilst minimising the risks.

2.10 It is vital in the context of CG, CSR and BRM that OSH professionals learn the language of the boardroom and are invited to participate in the CG/CSR/BRM processes. Reference is made in this Chapter to guidance pro-vided for OSH professionals by the Institution of Occupational Safety and Health (IOSH) in connection with BRM and OSH performance reporting in company Annual Reports. If this – and other – guidance is taken on board, it will enable OSH professionals to become more involved in – and add more value to – business risk management decision-making (see **Chapters 1 and 7**).

THE PRICE OF FAILURE

2.11 There are many disasters – catastrophes – that have occurred in the UK and elsewhere since the advent of the Health and Safety at Work etc Act 1974 and the concept of reasonable practicability, ie cost versus risk. In almost every case, management failure is cited as one of the root causes of these disasters. On closer examination, this management failure is sometimes as high up as board level. Decisions have largely been taken on a cost only basis, not a cost versus risk basis, and hence the wrong decision has been made based on incomplete data.

2.12 The Health and Safety at Work etc Act 1974, s 2(1) is the catch-all provision:

'It shall be the duty of every employer to ensure, so far as is reasonably practicable, the health, safety and welfare at work of all his employees.'

2.13 The term 'reasonably practicable' requires the risk to be weighed against the costs necessary to avert it. Costs in this context means not only financial (money) but time, people and trouble as well! If, compared with the costs involved, the risk is small (even trivial), then control measures need not be taken – ie it is not reasonably practicable to incur the cost. If, however, the risk is high and the cost of control is equivalent or lower, then it is reasonably practicable to incur cost. Only when the cost is grossly disproportionate to the risk, therefore, is it deemed to not be reasonably practicable.

2.14 In essence, this general duty implies the need for a risk assessment to be undertaken before a reasonably practicable decision may be satisfactorily made. This predates the explicit requirement for risk assessment by some 18 years – ie the Management of Health and Safety at Work Regulations 1992 (updated in 1999).

2.15 In a court of law, however, the burden of proof as to whether the action was reasonably practicable or not at a particular point in time rests with the employer – ie the duty holder, the controlling mind of the organisation, the director responsible for OSH management; ultimately the board itself!

2.16 The existence of a well-documented, suitable and sufficient, carefully considered risk assessment should go a long way towards supporting a case on what was or was not reasonably practicable, assuming the risk assessment and its associated, commensurate control measures were known, understood and adhered to by all concerned! As neither risks nor costs remain the same forever, what is reasonably practicable will change with time – hence the need to keep risk assessments, and their communication, recorded and up to date.

2.17 When things go wrong at work there is usually an investigation; in some cases, a formal enquiry takes place.

2.18   The following quotations have been taken from reports of enquiries into well-documented accidents, which resulted in serious loss of life. The enquiries found – and public opinion confirmed – that senior management failed when it came to OSH decision-making:

> 'A full investigation into the circumstances of the disaster leads inexorably to the conclusion that the underlying or cardinal faults lay higher up in the organisation. The board of Directors did not appreciate their responsibility for the safe management of their ships.'

Herald of Free Enterprise

> 'It is clear that the board did have proper regard to efficiency and economy: it is equally clear that they did not impose the same criteria when it came to the safety of operations.'

King's Cross Underground Fire

> 'The failure on the part of management to give proper and clear direction was a contributory cause of the disaster.'

Herald of Free Enterprise

> 'A concern for safety which is sincerely held and repeatedly expressed but, nevertheless, is not carried through into action, is as much protection from danger as no concern at all.'

Clapham Junction Railway Accident

2.19   The Health and Safety at Work etc Act 1974, s 37(1) is also highly relevant:

> 'Where an offence under any of the relevant statutory provisions committed by a body corporate is proved to have been committed with the consent or connivance of, or to have been attributable to any neglect on the part of, any director, manager, secretary or other similar officer of the body corporate or a person who was purporting to act in any such capacity, he as well as the body corporate shall be guilty of that offence ...'

2.20   If the impact of this section was fully understood – and enforced – at board level, then it is more likely that correct cost versus risk decisions would be taken, especially if the risk of individual prosecution and possible imprisonment was suitably and sufficiently assessed! Indeed the whole concept of corporate manslaughter and board level/directors' accountability is increasingly a vital component of good, visible corporate governance.

2.21   When things go wrong, the media has a good story. Bad news is good news; good news is no news (except for the 'And finally' section!).

2.22   Hence, from a BRM perspective, disaster avoidance is a good strategy to adopt in order to protect corporate reputation, brand values, good standing in the community, being seen to be 'green', being a good neighbour and all the other CG/CSR phrases.

2.23   The price of failure is much more than just the economic cost. There is vast humanitarian/social impact, which is invariably negative as far as the organisation is concerned. Any common or statute law cases resulting from such disasters will also have a negative impact on the organisation, both from a financial/insurance and media viewpoint.

2.24   Set out below are brief details of certain major disasters, which show how things can go wrong, and why. We have also identified the key lessons learned. In most cases, the major cause was the lack of a risk-based and effective OSH risk management system, integrated with normal day-to-day organisational and operational processes. Neither was OSH a consistent priority in all decision-making processes. Although the names of these disasters are relatively well-known, it is not always appreciated that each disaster could have been foreseen and avoided. The most significant causes relate to organisational and human factors and not technical failures, which require very different management approaches. We will return to this point in **Chapter 6**.

PIPER ALPHA

2.25   The exhaustive public enquiry brought out the detailed events of the disaster and also examined the background to the deficiencies that occurred on 6 July 1988. The report highlighted six major factors that contributed to the accident and the high loss of life (167 fatalities).

1   Initial explosion:
   - Caused by ignition of condensate leak.
   - Poor operation of PTW system (permit-to-work).
   - Lack of communication at shift changeover (1800 hrs).
2   Oil fire:
   - Breach of fire wall following initial explosion.
   - No consideration of condensate explosion: no explosion walls designed into structure; no HAZOP (hazard and operability) study at design stage.
   - Oil pipeline not shut down, thus providing additional fuel for the fire: not shut down on economic grounds, even though people were getting burned to death (decision changed one hour into the disaster).
3   Gas fire:
   - Neighbouring platform gas pipelines met on Piper Alpha where the oil fire was raging, thereby rupturing the gas pipelines and creating a ball of flame on Piper Alpha. This made escape extremely hazardous.

- Although this hazard was well understood by management (operational), no specific provision for this eventuality was made to protect the gas pipelines (eg fireproofing, water sprays).

4   Firefighting:

- Water deluge system fed via electrically driven water pumps which were put out of action when initial explosion destroyed the main power supply.
- Back-up diesel pumps should have automatically kicked in but these were switched to manual and controls could not be reached because of the oil fire.
- Diesel pumps switched to manual deliberately because divers were working near the sea-water suction intakes, a known hazard and a regular occurrence especially in the summer months during the night shift. Hence, for half the time the operability of the firefighting system was inhibited in an unnecessary and dangerous manner, as the diesel pumps were put on manual whenever divers were in the water, irrespective of whether they were near the suction intakes or not.
- Some of the spray heads were blocked because of the corrosive nature of sea water. This had been a known problem for four years prior to the disaster but a decision to replace the system with non-corrosive material had only recently been taken. At the time of the disaster, only part of the system had been replaced so it would only have been partially effective in any event, even if there had not been an explosion.

5   Accommodation block:

- In the event of emergency, those on shift assembled in the accommodation block together with those already there. The (rescue) helicopter pad was located on the roof of the accommodation block. Those who died in the accommodation block did so as a result of smoke/gas inhalation.
- The accommodation had some fire resistance built into it but it was not specifically designed to prevent smoke ingress. The majority of the smoke entered via fire-stop doors hooked open to facilitate escape. The emergency procedures should have ensured that fire doors were kept shut.
- As it should have been quickly apparent that it was too dangerous to land any rescue helicopters, the only means of escape was to jump into the sea and subsequently be picked up by one of the rescue vessels. However, no senior personnel gave any orders to evacuate in that way.
- Neither the installation manager nor any of his senior supervisors had received any thorough training on how to lead in a major emergency situation or had undergone any regular, simulated emergency exercises.

6   Safety auditing:

- Compliance with the permit to work (PTW) system was monitored each day and had also been audited by the parent company six months prior to the disaster. No deficiencies were reported.
- An annual fire safety audit report did not refer to the problem of blocked deluge heads.
- A previous audit report recommendation only to put the diesel fire pumps on manual when divers were working in close proximity to the pump intakes had never been implemented.
- There was not a shortage of audits but a definite shortage of quality auditing and recommendation(s) follow-up. The absence of critical comment in audit reports lulled senior and platform management into a false sense of security, and safety!

2.26 From the above six major contributory factors, the following lessons became apparent:

- There must be a systematic risk assessment of all potential major hazards at the design stage.
- Interactive effects in emergencies between linked operating units must be thought through via techniques such as HAZOP.
- A system for the timely resolution of faults in safety-critical equipment must be part of normal operating management.
- The PTW system needed reviewing and re-communicating.
- All the deficiencies listed in the six categories were the responsibility of senior/line management, from the design stage right through the operating life of the platform and its associated equipment.
- A systematic approach was required. The Piper Alpha deficiencies clearly illustrated failures in a number of systems. Either there was a system but it was inadequately designed and executed, eg the PTW, or there was no system in place where one should have existed, eg the lack of a systematic method for assessing major hazards.
- Quality of safety management was critical. All operations should have known and well understood safety systems in place that meet the defined requirement each and every time on each and every day.
- Auditing is vital as part of any organisation's OSH management system. It is important that quality auditing by competent internal and external auditors is regularly in place and that audit recommendations are implemented and followed up. Individual responsibilities and timescales should be firmly fixed in audit reports. All reports should be communicated to all concerned so as to ensure findings are acted upon.

HERALD OF FREE ENTERPRISE

2.27   On 6 March 1987, the cross-channel roll-on/roll-off ('ro-ro') ferry, *Herald of Free Enterprise*, sank soon after leaving Zeebrugge with the loss of 186 passengers and crew. The vessel sank because the large inner and outer bow doors through which vehicles roll on and off had been left open, thereby allowing sea water to enter the ship's vehicle decks. The surge of water moved to one side of the ship causing it to roll onto its side.

2.28   The official report into the disaster considered the causes of the accident, the deficiencies in design and the failure of management within the organisation.

- The assistant bosun – who should have closed the bow doors – was asleep and did not hear the Tannoy announcement that the ship was about to sail.
- The ferry company changed from doors that open upwards (known as visor doors) to doors that swing open vertically (known as clam doors). The visor doors could clearly be seen – when open – from the bridge, but the clam doors could not be seen. No visible warning device had been fitted to indicate to those on the bridge that the doors were open or closed.
- On numerous occasions other ships had sailed with the bow doors open, thus prompting the captains to ask for such devices – indicator lights – to be fitted. These requests were ignored – and even treated with contempt – by head office management.
- The bosun did not ensure that his assistant shut the doors, even though he was the last to leave the immediate vicinity of the bow doors.
- An instruction that the officer loading the vehicle deck should ensure that the bow doors were secure when leaving port was ignored.
- The assistant bosun did not have to report to the bridge that the doors had been shut, only if there had been a problem. No report, no problem – management by exception!
- No bulkheads were designed into ro-ro ferries to make vehicle movement – and also the ingress of water – easier.
- The board of Directors did not appreciate their responsibility for the safe management of the ships. From top to bottom the body corporate was infected with the disease of sloppiness.
- Responsibilities at board level were not clear.
- There was no system of auditing the standard of operation, going against a 1986 Department of Transport recommendation.
- There was pressure for the ship to leave as soon as possible so as to enable a fast turnaround. Quicker turnarounds meant more crossings per day, a decision taken at board level.
- Written instructions for the ship's officers were unclear and contra-

dictory. Although this had been pointed out in 1982, it had been ignored.

2.29 As with Piper Alpha, management leadership failings and management system failures – both on the ship and on shore – led to the *Herald of Free Enterprise* disaster. Warning indicator lights were fitted to all other ships in the fleet within days of the *Herald of Free Enterprise* sinking.

CLAPHAM JUNCTION

2.30 On 12 December 1988, a commuter train crashed head on into the rear of a stationary train near Clapham Junction Station. The moving train then struck a third train coming in the opposite direction. In all, 35 people were killed and almost 500 injured.

2.31 The stationary train had stopped because a signal suddenly changed from green to red. The driver stopped at the next signal to report the fault by phone to the signalman. Unfortunately, the fault in the signal was that it showed green when it should have shown red. Hence the commuter train's driver thought it was safe to proceed and smashed into the stationary train and subsequently veered into the oncoming train.

2.32 The official report into the crash considered the immediate and underlying causes of the accident, the vast majority of which are due to deficiencies in management systems and many types of human error.

- During the two weekends before the accident, a new signalling system had been installed, during which one of the signalling technicians failed to follow agreed procedures. After disconnecting a redundant wire, he failed to cut it back to prevent inadvertent reconnection. He also used old – instead of new – insulating tape to cover the bare end.
- This failure to adhere to agreed procedures had been going on for 16 years and had not been picked up by supervision. The errors were genuine mistakes resulting from poor training and supervision and a lack of written instructions.
- There was no independent check of the technician's work. There was no monitoring system to flag up when retraining may be needed.
- The technician had been working seven days a week for many weeks.
- The supervisor did not check the work because he was busy leading a team of contractors and was also carrying out signalling work himself because of the pressure to get the job completed in good time. He had no time left to be an efficient supervisor!
- Neither the supervisor nor the technician had seen two sets of written instructions concerning wire checks/counts issued in 1985 and 1987. These instructions had been posted to them and it was taken for granted by the issuing management that they had received, read, understood and implemented them!

- No checks were made to see if the instructions were being followed.
- The acting tester did not do a wire check/count, only a functional test, as he understood that was all that was required. He had only been a tester for a few months and had no induction training.
- The manager in charge of the supervisor turned a blind eye to irregularities as he was due to retire in 18 months and did not wish to interfere with the accepted way of working.
- Senior management were aware of failings in the testing system. Recommendations had been made for improvements but these had not been implemented or checked on.
- Five wiring failures similar to those at Clapham Junction occurred elsewhere during 1985 without causing a serious accident. The only result from what can only be described as a series of near misses – warning signs – was a new provisional instruction posted out to managers, supervisors and technicians. There was no get-together, briefing session, training course or toolbox talk! Hence the new instruction was never properly explained, understood or implemented, so no change occurred to the existing working practices.

'A concern for safety which is sincerely held and repeatedly expressed [by British Rail] but nevertheless is not carried through into action [by BR management] is as much protection from danger as no concern at all.'

2.33   The above quotation from the official report sums up the attitude of the then-BR senior management – paying lip service to safety but not being protective in terms of identifying hazards, assessing risks, implementing commensurate controls to reduce or eliminate the risk, and ensuring that the controls are put in place – ie looking at both sides of the cost versus risk equation.

2.34   The official report recommended a TQM (total quality management) approach, a review by outside consultants and independent auditing – all component parts of the BRM process.

THE DEVELOPMENT OF CORPORATE GOVERNANCE (CG)

2.35   Partially as a result of the spate of corporate disasters in the 1980s – some of which have been discussed above – and partially because of the desire from certain organisations to be 'seen to be green' and to 'be a good neighbour', the need for more non-financial corporate accountability became the flavour of the decade in the 1990s.

2.36   Indeed, there has been increasing scrutiny of the responsibilities that organisational managements have towards all their stakeholders – shareholders, suppliers, employees, contractors, customers and the general public at large – and also their obligations to the local communities in which they operate, and to society, of which they and their stakeholders are part: hence corporate governance (CG) and corporate social responsibility (CSR).

2.37   In the UK, a series of committees were commissioned during the 1990s to report on and advise as to procedures designed to improve the competence of executive directors, the manner in which they should discharge their duties, and the need for additional reporting and supervision.

Cadbury

2.38   This report was produced in December 1992, following the appointment of a House of Commons Select Committee chaired by Sir Adrian Cadbury, with the aim of conducting an in-depth review of corporate governance. The outcome was a set of proposals on corporate governance and a code of best practice. Specifically, the report called for the introduction of a level of uniformity in the manner and conduct of board activities. The Code required all companies to adhere to the proposals.

2.39   In June 1993, the London Stock Exchange announced its requirement for all publicly listed companies to adopt the Code and include a compliance statement in their annual reports.

2.40   The Code required:

- The board to report on the business as a going concern.
- The board to report on the effectiveness of the system of internal control.

2.41   It focused almost entirely on financial concerns within organisations, but did make a few references to risk management.

2.42   However, the enactment of the Code was flawed, as the very people who were being asked to report on the adequacy of internal control were those who had been given the responsibility for installing it! Having said that, there were some useful specific recommendations, which have contributed towards the overall goal of good corporate governance through what amounts to risk management measures. These include:

- The appointment of non-executive directors.
- Procedures for directors to take independent professional advice (eg legal, financial).
- Establishment of an independently mindful audit committee.
- Comments on directors' service contracts.

2.43   The response of organisations to Cadbury largely focused on the system of internal financial controls. In some organisations, this resulted in the formation of a risk management committee, or advisory group, usually chaired by the Financial Director, and including representatives from auditing, treasury and risk management.

Hampel

2.44   This report was produced in January 1998, following the setting up of a Committee on Corporate Governance under the chairmanship of Sir Ronald Hampel. This committee – and its report – focused on the promotion of better standards of corporate supervision and control at board level. The recommendations of the report were incorporated into the Combined Code endorsed by the London Stock Exchange in June 1998.

2.45   This Combined Code was, in effect, a set of CG principles that incorporated concepts from Cadbury (see above, **para 2.38**), Greenbury (the report on directors' remuneration) and the work of the Hampel committee.

2.46   Hampel recommended that organisations comply (how they were to comply and were seen to comply is not prescribed) with certain guiding principles, some of which have risk management implications:

- Principle A4: the board should be supplied with information to enable it to discharge its duties.
- Principle D2: The board should maintain a sound system of internal control to safeguard shareholders' investment and the company's assets. (This covers not only financial controls but operational and compliance controls and hence risk management.)

Turnbull

2.47   We first introduced this report in **Chapter 1**. The report mentioned that the management of OSH risks should be included in organisations' main risk-management processes. The report was produced in September 1999, under the auspices of the Institute of Chartered Accountants of England and Wales (ICAEW) working party chaired by Nigel Turnbull. The Turnbull/ICAEW report was given the backing of the London Stock Exchange as it produced guidance on the Combined Code (see Hampel, above, **para 2.44**) and the basis upon which compliance should be audited.

2.48   Primarily the guidance required organisations to identify, in a formal way, their significant risks – both financial and non-financial – and how they managed them. The organisation's Annual Report should thereafter include commentary on the identified significant risks and their commensurate control measures.

2.49   The Turnbull guidance therefore has resulted in a greater emphasis being placed on risk identification, assessment and control at board/director level for all risks, operational as well as financial.

2.50   It requires the board to consider proactively the following business aspects:

- The nature and extent of the risks facing the organisation.
- The extent and categories of risk which it regards as acceptable for the organisation to bear.
- The likelihood of the risks identified actually materialising.
- The organisation's ability to reduce the incidence and impact of those business risks that do materialise.
- The costs of operating control measures relative to the benefit (ie cost versus risk) thereby obtained in managing those significant risks.

2.51   Hence the Turnbull guidance requires organisational boards to:

- Have a defined process for the review of effectiveness of internal control.
- Review regular reports on internal control.
- Consider key risks and how they are managed.
- Check the adequacy of actions taken to reduce/eliminate business risks.
- Consider the adequacy of monitoring.
- Conduct an annual assessment of risks and the effectiveness of internal controls.
- Make a statement on this process in the Annual Report.

2.52   Specifically, Principle D2.1 requires that:

'The directors, at least annually, conduct a review of the effectiveness of the organisation's system of internal control and should report to shareholders that they have done so. This review should cover all controls, including financial, operational and compliance controls and risk management.'

2.53   When reviewing reports, the board should:

- Consider the significant risks and assess how they have been identified, evaluated and managed.
- Assess the effectiveness of the related system of internal control in managing the significant risks – having particular regard to any significant failings or weaknesses that have been reported.
- Consider whether necessary control actions are being taken promptly to remedy the highlighted failings/weaknesses.
- Consider whether the findings indicate the need for a more extensive internal controls monitoring system.

2.54   In order to demonstrate Turnbull compliance, it is necessary for organisational boards to:

- Have the whole management reporting system embedded into the culture of the organisation – built in, not bolted on!
- Ensure the reporting system is common to all business functions, reporting dates, etc.

- Undertake an annual assessment for the purposes of making an annual CG statement in the Annual Report.
- Consider any changes since the last review in the nature and extent of significant risks and the organisation's ability to respond effectively to changes in its internal and external business environment.
- Review the organisation's business and operational management structure to identify changes that might alter its risk appetite and/or profile.
- Review the adequacy and timeliness of the risk-monitoring procedures in place.
- Review how the organisation manages the residual/tolerable/acceptable risk.

2.55   All the above can be seen to apply to the management of the OSH risks within an organisation.

2.56   The Turnbull guidance recommends that, although the board should set the strategy/tone for BRM, management should be made accountable to the board for monitoring the system of internal control (the embedded system) and for providing the board with an assurance that it has done so. Furthermore, all employees should be made to have some responsibility for aspects of internal control as part of their accountability for achieving their current year objectives (CYOs).

2.57   They, collectively, should be given the knowledge, skills, information and authority to establish, operate and monitor the system of internal control. This should manifest itself through job descriptions, CYOs, key performance indicators (KPIs), action planning, and regular formal progress reporting up and down the management hierarchy.

Higgs

2.58   This Report was published in January 2003, following a request by the UK government to examine CG in more detail, which led to the commissioning of the Higgs Report. The report recommended some amendments to the original Combined Code (see above) and provided guidelines for the roles of non-executive directors and chairpersons of organisations.

2.59   Some of the key guidelines are as follows:

- The number of meetings of the board and its main committees should be stated in the Annual Report, together with the actual attendance of the individual directors.
- At least half the board – excluding the chairperson – should be independent, non-executive directors.
- Non-executive directors should meet as a group at least once per year without any executive directors – including the chairperson, if execu-

tive – present. The Annual Report should identify whether such meetings have taken place.

- Appointments to the board should be made via a nominations committee.
- The board should declare to the shareholders the reasons in support of the appointment of an individual as a non-executive director.
- Full-time executive directors should not take on more than one non-executive directorship nor take the chair of a major company/organisation.
- Board performance – its committees and individual members – should be evaluated at least once per year.

2.60   The Higgs Report also led to the Tyson Report which explored routes by which organisations might comply with the Higgs recommendation that non-executive directors should be drawn from a wider range of backgrounds and experience, than has tended to be the norm up until the last two years or so. In other words, to do away with the non-executive directors' club!

Sarbanes–Oxley

2.61   This piece of USA legislation, the Sarbanes–Oxley Act 2002, set out requirements for corporate compliance and in-depth auditing of operational standards and controls. It is geared to restoring investor confidence in the wake of the Enron and WorldCom affairs. In some respects the Act – which came into force at the end of July 2002 and had to be implemented by the Securities and Exchange Commission (SEC) by 29 August 2002 – goes further than UK reporting practice on internal control.

2.62   Indeed, those UK companies that have a secondary listing in the USA have had to put in place revised processes that allow their CEO (Chief Executive Officer) and CFO (Chief Financial Officer) to personally report on the robustness and efficacy of the internal control system. In addition, the SEC has also published guidance on CG standards for organisations whose shares are quoted in the USA.

2.63   The Sarbanes–Oxley Act also imposes additional duties on the appointed external auditors of the organisation, which will increasingly be discharged via the use of non-financial expertise.

Nicholson

2.64   This report was published in July 2003. It was prompted by the Financial Reporting Council (FRC) asking Sir Bryan Nicholson to chair a committee charged with bringing together the then-current Combined Code, Turnbull and Higgs reports, and re-issuing them under one all-embracing corporate framework. The resulting report was entitled *The Combined Code on Corporate Governance*.

2.65   The Nicholson Code follows the structure of the Hampel Code in that it describes important main principles of good CG together with new supporting principles.

2.66   The Code states that an effective board should head every organisation, which should be collectively responsible for the organisation's success. The new Code contains a provision that the organisation should arrange appropriate insurance cover for directors.

2.67   The Nicholson Code places great importance on the make-up of the board, stating that at least half the board members should be non-executive/independent. The report also highlights the need for all directors to regularly update their skills and knowledge on CG matters.

2.68   The Code came into effect on 1 November 2003 and the main changes have been to board practices/appointments and the role of the audit committee. Nicholson has included much more detailed public reporting requirements, which have greatly assisted the analysis of the CG status of an organisation, thereby allowing stakeholders/shareholders/ potential investors, etc, to judge just how committed the board is to effective CG and the management of risk, including OSH and environmental concerns.

2.69   Copies of the Nicholson Combined Code are freely available as a download from the FRC's website: www.frc.org.uk.

Operating and financial review

2.70   Published in May 2004, this Government Consultation Document was entitled *Draft Regulations on the Operating and Financial Review (OFR) and Directors Report*. This document was in response to work of the 2003 Taskforce Accounting for People – sponsored by the UK DTI (Department of Trade and Industry) – and also the recommendations of the independent Company Law Review and the Accounts Modernisation Directive of the European Union.

2.71   This document states (Para 2.5):

'Directors need to decide in good faith what would be most likely to promote the success of the company, taking account of a wide range of factors, within and outside the company, which are relevant to achieving its objectives and to an assessment of the business. These factors may well include the company's impact on the environment and on the wider community, and on its relationship with employees, customers and suppliers.'

2.72 These OFR regulations – once approved – will become mandatory during 2005, currently 1 April.

2.73 There is a clear implication here that the management of environmental – as well as health and safety – risks are key in the CG process. However, a recent (September 2003) poll undertaken by IT company, Unisys, of 400 UK businesses found that almost 34% of organisations have still not implemented some of the Cadbury, Turnbull and Higgs recommendations, citing lack of time and resources! Almost 50% believe CG restricts entrepreneurial spirit, in spite of increasing pressure placed on organisations to gain accreditations and follow specific guidelines in the light of recent corporate disasters (eg Enron, WorldCom).

2.74 The survey found that CG is largely driven by customer need (82%) and pressure from shareholders (77%). However, 70% of those surveyed did not think that meeting CG guidelines improved their competitiveness, and 75% thought that accreditation did not improve their profile. The general attitude of those organisations surveyed was that the regulatory processes put into place for good CG were viewed as restrictive 'tick the box' exercises that stifle creativity and agility!

2.75 We return to the subject of OFR in **Chapter 7**.

CORPORATE SOCIAL RESPONSIBILITY

2.76 Corporate social responsibility (CSR) is an important part of the overall approach to CG. Increased awareness of society towards CG and, more especially, corporate behaviour and the societal impacts of outright failure (poor CG) on local and national communities has gathered apace in the last ten years or so (eg Bhopal). This has more recently manifested itself in a genuine concern that big business seemed to have power greater than that of elected governments!

2.77 Whatever causes the ultimate collapse of a business/organisation, it is the stakeholders – shareholders, employees, suppliers – that lose out. Those organisations that have failed in recent times have ultimately been destroyed because of the loss of their good reputations.

2.78 Hence CSR is very important as a core value or core business concept in protecting the organisational reputation, brand, image, etc, and thereby protecting the bottom line. It should therefore be seen as a business risk management tool rather than a drain on (short-term) financial resources. Getting it wrong may well be much more expensive than getting it right! We are again in a cost versus risk situation!

2.79 Large, multinational companies that are household names are assumed to meet high standards of CSR – whether they do or not. When they are shown to fall short of society's expectation, it is then that insurmountable difficulties arise and they find themselves on the slippery slope to business oblivion.

2.80 This is where CSR becomes an excellent way to protect certain organisational assets, especially brand value. The brand carries the reputation of the organisation and embodies customer expectations, including concepts of quality, environmental friendliness, human rights and OSH. If an organisation has a poor OSH record, their ability to market products and services which are advertised as 'healthy' and 'safe' is severely jeopardised.

2.81 The household name product/service is believed by the customer to embody all their perceived expectations in many different areas of business including those mentioned above and also:

- Business ethics.
- Employment practices.
- Being truthful in company reports/advertisements/product recall notices.
- Being friendly to the communities in which it operates – being a good neighbour.
- Believes in equal rights.
- Opposes tyranny.
- Cannot be bribed.
- Uses only the finest, safest, healthy ingredients from legitimate, sustainable sources.
- Does not exploit children/young persons in the manufacture, distribution and marketing of its products/services.

2.82 As a result of the above, CSR has developed at an astonishing rate in recent years, culminating in a new-look government CSR website at: www.societyandbusiness.gov.uk and is now seen to be 'good business'.

2.83 According to Business in the Community (BiTC) – who produce and publish the Corporate Responsibility Index (CRI) – many investment analysts believe that an organisation that proactively follows a CSR route is much more attractive to investors than one which does not. Indeed this has, in recent times, led to the formation of the FTSE4 Good index for socially responsible investing. Its selection criteria are:

- Upholding and supporting human rights.
- Environmental sustainability.
- Social issues, including OSH.
- Positive relationships with stakeholders including investors, employees, customers, suppliers, insurers, trades unions, government and non-government organisations, enforcers/local authorities, other businesses, the media, and local communities.

2.84   BiTC/CRI was the first objective CSR measure within the UK. BiTC defines CSR as an organisation's positive impact on society and the environment through its operations, products or services and through its interaction with key stakeholders such as those listed in the paragraph above. The BiTC/CRI survey now goes out annually to hundreds of FTSE and other leading companies and is rapidly becoming an agenda item in UK boardrooms. The survey covers a range of indicators including OSH and environmental issues. These indicators are grouped into four main sections:

1   Corporate strategy.
2   Integration.
3   Management practice.
4   Performance and impact.

2.85   Corporate strategy includes establishing CSR values, goals, leadership, and stakeholder identification, policies and risk management.

2.86   Integration involves incorporating the CSR strategy into the normal way of doing business – as with Turnbull, embedded into the organisational culture and framework. It should rank alongside and include: overall business conduct/ethics, performance management – targets, pay and bonuses, decision-making, training/career development, stakeholder involvement/ engagement, and risk reporting.

2.87   Management practice encompasses areas such as: community, environment, workplace (including OSH) and the marketplace. In each of these four areas there is a need to consider: identification of CSR issues/ concerns; impacts/risks of decisions, objectives and targets; responsibilities/ accountabilities; employee involvement and communication; training; communication with all external stakeholders; management systems; supplier audits; and product stewardship.

2.88   Performance and impact, where respondents select a minimum of two environmental impact areas and two social impact areas. A total of six impact areas are to be completed. The selection should ideally be based on the six impacts that are most significant to the organisation. At least two out of the five social impact areas should be selected; these are:

●   Product safety.
●   OSH.
●   Human rights in the supply chain management.
●   Workplace diversity.
●   Community investment.

2.89   All organisations are asked to report on two environmental impacts, namely global warming (energy and transport) and waste management. The BiTC/CRI scoring system for each section is a maximum of 22.5, giving a total of 90 over the four sections. The remaining ten marks may be awarded for an additional fifth section labelled 'Assurance'. In addition to the completion of the survey, there is also a need for supporting evidence throughout.

2.90   BiTC's second CRI was published in the *Sunday Times* on 14 March 2004 and, for the first time, the top 100 'Companies that Count' were listed. A large number of the companies (70) included OSH in their selected impacts.

2.91   The top ten were as follows:

1   National Grid/Transco.
2   BP.
3   Unilever.
4   Veolia Water UK.
5   Aviva.
6   Waste Recycling Group.
7   Co-operative Bank.
8   J Sainsbury.
9   Barclays.
10   Rolls-Royce.

2.92   The full list is freely available from BiTC (telephone: 0870 600 2482) or via their website at www.bitc.org.uk.

2.93   The BiTC OSH-related questions are linked to the operation of OSH management systems. The questions look for aspects such as:

● Evidence of a structured risk management process.
● Identification of key OSH/business risks.
● Clear OSH management responsibilities and accountabilities.
● Key OSH performance indicators and targets.
● Systems to monitor OSH product performance.

2.94   Another guide to good CSR is the Social Accountability Standard, SA8000: 2001 (Social Accountability International, New York). This Standard should greatly assist in the process of identifying and prioritising corporate risks and their impacts on the business. This risk assessment – and commensurate control – is fundamental in the development of a cogent CSR strategy.

2.95   As with BS 8800: 1996 (revised during 2004) 'Guide to Occupational Health and Safety Management Systems' (see **Chapter 3** of this book), the SA8000 includes an initial, baseline audit followed up by surveillance and certification/re-certification audits in order to achieve continual improvement via the implementation of a programme of corrective actions. The subject areas included in the SA8000 Standard are:

● Child labour.
● Forced labour.

- OSH.
- Freedom of association.
- Right to collective bargaining.
- Discrimination (of any form).
- Disciplinary procedures.
- Working hours.
- Remuneration.
- Management systems.

2.96   So it is clear that OSH concerns figure prominently in both the CG and CSR agendas. Hence it is vitally important that OSH professionals take this on board and learn the language, strategies and tactics of the boardroom!

OSH AND BUSINESS RISK MANAGEMENT

2.97   At the recent (May 2004) RoSPA Conference and Exhibition at the NEC, Birmingham, Jane Kennedy, the current Minister of Work with responsibility for health and safety, was reported as saying that risk management is the key to reducing the number of deaths and injuries in the workplace but that this can only happen if everybody works together and adopts a brave new approach to health and safety.

2.98   She highlighted the fact that the costs of poor health and safety to business and society at large are enormous and the level of individual suffering is unacceptable. Although Britain has a record to be proud of, the workplace and its hazards are changing. She said that UK plc has to respond to this change and look towards more innovative ways of working to tackle new issues – such as stress and musculoskeletal disorders – if we are to make further inroads into health and safety improvements. She concluded by saying that the government goal was not a risk-free society but one where risk was properly managed and understood.

2.99   This clear directive towards BRM gives OSH professionals a golden opportunity to revitalise the OSH and BRM programmes within their organisations.

Overview of business risk management

2.100   Increasingly, organisations are taking a holistic approach to the management of risk which requires OSH professionals to be in tune with the principles, language and practices of BRM. Those who can understand and apply its principles – and can communicate with other business disciplines – have the opportunity to be more efficient and effective. Well-informed OSH professionals are better able to make the business case for the management of OSH risk within the wider context of organisational and operational risk.

2.101   BRM is a strategic process which aids and supports decision-making at both strategic and tactical levels within an organisation. An improved understanding and hence management of all risks likely to positively affect the organisation will lead to better performance and competitive advantage, especially when hazards and threats are identified – and the risks assessed and controlled – in the same manner as opportunities and rewards.

2.102   BRM may therefore be defined as the eradication or minimisation of the negative effects of pure and speculative risks to which an organisation is exposed in order to maximise the positive impact on the bottom line. It should be noted that pure risks could only result in a loss to the organisation – eg injury, disease, damage or death. Speculative risks may result in either gain or loss – ie speculate to accumulate.

2.103   Hence BRM is used within organisations to:

- HConsider the impacts of foreseeable significant risky events on the performance of the organisation.
- Respond appropriately to internal and external changes in risk perception.
- Devise strategic options for eliminating or controlling all significant risks and their impacts.
- Link these options to the general decision-making and control framework used by the organisation.

2.104   The requirement for the BRM approach is given added credibility as a result of the Turnbull Guidance, which requires UK stock-market-listed organisations to identify, record and manage their significant risks in a suitable manner. Systems for regular review of risks and amendment of internal controls must be embedded within the organisation's management framework, together with statements in company annual reports confirming the effectiveness of these systems.

2.105   OSH risks and controls should be included when and where they present significant operational and compliance risks, which may well be associated with wider financial and reputational losses.

2.106   Effective internal controls safeguard stakeholder/shareholder investment and organisation assets (manpower, materials, machinery, methods, manufactured goods/services, money – the six Ms) especially when they are risk-based and embedded into the organisation's management systems – built in, not bolted on! The growth of corporate governance and socially responsible investment indices – such as FTSE 4 Good – allows potential investors to choose organisations that demonstrate good corporate social responsibility. This should automatically include aspects of OSH and environmental performance.

The holistic approach to business risk management

2.107   OSH management is clearly seen as an important part of BRM, and the growth/evolution of BRM may be traced back to the pure risk management of negative impacts on people, property and products – the conventional approach to occupational safety in the late 1940s and 1950s.

2.108   During the 1960s and 1970s, the advent of loss control expanded into the management of the pure risks associated with areas such as fire, security, environment and business interruption.

2.109   Following on in the 1980s and 1990s, the whole concept of the management of both pure and speculative risks gathered apace with the expansion into areas such as financial/insurance implications, internal audit, brand/reputational risk, IT/e-commerce, and business continuity.

2.110   The 1990s/2000s era has resulted in the holistic/corporate governance scenario, which now encompasses:

- Management systems.
- Risk assessment and controls.
- Turnbull guidance compliance.
- Corporate social responsibility (CSR).
- Socially responsible investing (SRI).
- Risk reporting in annual company reports.

2.111   Hence, business risks should be treated in totality, because of the impact of one risk on another, rather than via the conventional, piecemeal or compartmentalised approach – sometimes referred to as silo management. The watchwords in the holistic approach to BRM are therefore: mind the gaps! This is discussed in more detail in **Chapter 7**.

2.112   It is vitally important to recognise these interrelationships and impacts – both positive and negative – of the various types of pure and speculative risk. Hence, the management of OSH risks should never be treated in isolation, especially when consideration is given – hopefully at board level – to the negative impact that poor OSH management can have on other business risks such as brand/reputation, insurance, business continuity and financial well-being.

2.113   This is one reason why organisations integrate their OSH management systems with those used to manage environmental impacts or quality, other key factors that affect reputation – and hence the ultimate success – of the organisation (see **Chapter 3**).

2.114   The bringing together of insurance (risk transfer) and loss control (risk reduction) in the 1980s/1990s was the final stage in the evolution of holistic BRM and clearly indicates the value and positioning of OSH professionals in the drive to get organisations to adopt, adapt and continually improve their BRM processes.

Key elements of the BRM process

2.115   The BRM process comprises identification, evaluation and control, plus the monitoring, audit and review stages common to all effective management systems.

2.116   Risk identification (typically referred to in the OSH context as hazard identification) may be achieved via a multiplicity of techniques, most of which are well known to the OSH profession. These include:

- Application of/compliance with standards.
- Inspections and audits utilising bespoke checklists.
- Workforce involvement, consultation and participation.
- Accident/loss investigations.
- Job/task analysis.
- Scenario planning.
- Stakeholder consultation.

2.117   Organisations having mature OSH management systems but limited experience of wider BRM issues can adapt their existing OSH processes and recording formats to cover other key risk areas – fire, security, environment, products, business continuity – highlighted above. OSH professionals can contribute to analytical techniques such as SWOT (strengths, weaknesses, opportunities and threats) and PEST (political, economic, social and technological), which business managers within the organisation may already use.

2.118   Risk evaluation (or measurement) is a technique designed to quantify the risk in terms of its economic, social and/or legal impact on individuals and/or the organisation. This will include statistical aspects such as probability, frequency and severity of occurrence and is similar in nature to the legal requirement for OSH risk assessments. Indeed, it has been argued that risk assessment is, in essence, a combination of the identification and evaluation elements of BRM.

2.119   Risk-control strategies may be classed into four main areas:

1   Avoidance.
2   Retention.
3   Transfer.
4   Reduction.

2.120   Risk avoidance involves conscious decision-making by an organisation to completely avoid a particular risky process or event by discontinuing the operation producing that risk. This presupposes that a complete risk analysis has been undertaken in order to provide the organisation with its risk profile. Any risks assessed in the very high category may well have to be discontinued because of their adverse economic and/or social and/or legal impact on the organisation and its reputation.

2.121   Risk retention involves the risk being managed and financed within the organisation. Hence, any loss arising as a result of poor risk management will adversely affect the bottom line. This in itself provides a degree of economic motivation towards effective BRM! Risk retention may be subdivided into two risk subsets: risk retention with knowledge (where a full analysis, assessment and profile is in place) and risk retention without knowledge (where nothing proactive has been done and any risk management is purely reactive – ie after the event).

2.122   Risk transfer involves the legal assignment of the costs of certain potential losses from one party to another via the use of some form of contractual agreement. The most common method of affecting such transfer is via the use of individual or commercial insurance – eg motor, household, property, fire, products, employers' liability, etc. The insured (client) pays a risk transfer payment (the insurance premium) to the insurance company. As and when an insured loss occurs, the insurance company pays compensation to cover the cost of the loss to the insured, assuming the policy terms and conditions are met. What the insured does with the compensation payment is entirely up to the individual or organisation concerned. Other forms of contractual risk transfer include sales contracts and employment of third parties (eg contractors). In certain situations, very high-risk activities may be transferred from one organisation to another or indeed from one part of the world to another!

2.123   Risk reduction involves the ongoing management of (residual/ acceptable) risk within the organisation via the effective implementation of a programme designed to protect the organisation's assets (remember the six Ms – above) from wastage caused by accidental loss. The component parts of such a risk reduction or loss control programme include subject areas such as:

- Occupational safety.
- Damage control – prevention of damage to property, products, equipment, motor fleet, etc.
- Fire prevention.
- Security/anti-fraud measures.
- Computer/IT protection.
- Occupational health/hygiene.
- Environmental protection.
- Product or service safety/quality assurance.
- Public safety/liability.
- Business continuity.

2.124   This is where the holistic approach is vital, especially as many of the above risks and their commensurate control measures are both interdependent and interactive. This is definitely the time to mind the gaps and not be operating in silos or compartments.

2.125   A prime example of this may involve a decision to lock off fire doors during work periods in order to reduce the security risk of petty pilfering and/or vandalism. In the event of a fire, a low risk rapidly escalates into a very high – and totally unacceptable – risk!

2.126   To complete the management system cycle, and in order to move towards the goal of continual improvement, the BRM process also involves monitoring, audit and review, as is the case with other systems such as those for OSH, environment and quality.

2.127   All risk-control systems should be proactively and reactively monitored in order to ensure that they are efficient, effective and are working as designed. Any rectification needed should be actioned within finite timescales by named responsible individual managers/supervisors.

2.128   The BRM process should be audited – at least annually – to ensure that both strengths and weaknesses are equally highlighted and action taken where necessary to ensure continual improvement.

2.129   The whole process should also be reviewed at board level – again at least annually – to enable the identification, evaluation and control components to be kept up to date, especially in times of significant change within the organisation. This will also enable accurate risk reporting to be featured in Company Annual Reports.

Institution of Occupational Safety and Health (IOSH) – position statement on business risk management

2.130   In its 2002 document on BRM, www.iosh.co.uk/technical, IOSH states that, as the OSH professional body, it believes that it is important for OSH professionals to build links and co-operate with others involved in the BRM process. This may include using tools and competencies originally developed to support good OSH management, adopted, adapted and improved to add value in the wider BRM context. IOSH encourages its members to develop an understanding of the language and tools used by business managers and to take every opportunity to ensure that significant OSH and environmental risks are firmly on the BRM agenda.

REPORTING OSH PERFORMANCE IN COMPANY ANNUAL REPORTS

2.131   The provision of occupational safety and health performance information in company Annual Reports is currently a growth area since the advent of the Turnbull Guidance on internal control in 1999.

2.132   Turnbull stressed the need for organisations to manage their non-financial risks – including occupational safety and health and environmental concerns – with the same rigour they apply to financial risks. Indeed, all business risks – both pure and speculative – should be identified, assessed and controlled via a process, which is embedded into the organisation's management systems. A key issue raised by Turnbull was the need for Stock Exchange listed companies to publicly report in their Annual Reports on progress made on corporate governance and corporate social responsibility issues, including occupational safety and health and environmental matters.

2.133 The whole concept of risk – including health and safety – reporting is gathering pace. Organisations are beginning to apply the same rigour to non-financial, operational reporting as they have done for financial reporting over many years.

2.134 The government enforcers of risk management have been monitoring the output of relevant performance data over the last few years and are actively encouraging companies to be more open and honest in their Annual Reports.

2.135 For example, the Environment Agency (EA) recently slated organisations for whitewashing their environmental records. An EA study found that 89% of Stock-Market-listed companies included basic environmental (green) information. However, closer examination of these disclosures revealed that the vast majority lacked depth, rigour and quantification. Indeed, only 10% of company Annual Reports gave information on topics such as how they deal with waste or how their work might affect climate change.

The Health and Safety Commission/Health and Safety Executive approach

2.136 The Health and Safety Commission/Executive (HSC/HSE) have also been monitoring the output of occupational safety and health performance data in company Annual Reports.

2.137 Indeed, Action Point No. 2 of the Revitalising Health and Safety Strategy (RHSS) recommends that large businesses report publicly to a common standard on health and safety issues by the end of 2002.

2.138 Their initial study in 2000 concerning the Annual Reports of the FTSE top 350 companies revealed of those that actually responded to the HSE's request for information (227 out of 350 – 64.9%) only 107 (47.1%) contained any occupational safety and health information. The survey report stated that there was a great variation in the quality – and quantity – of the information published.

2.139 As a follow-up exercise, the HSC challenged all 350 to improve their reporting by including relevant occupational safety and health performance information in their 2002/2003 reports. It is highly likely that HSE will be closely monitoring future reports with a view to extending the request to cover all listed companies and public bodies. Eventually the target will be for all organisations having more the 250 employees to include relevant occupational safety and health performance data in their Annual Reports by the end of 2004.

2.140 More recent research published in 2003 showed that the figure including health and safety information in Annual Reports had risen from 47.1% up to 78.0%. In order to keep the figures increasing, the HSE and Local Authorities (where they are the lead authority) have set up top-level meetings with over 70 companies to explain what is required and to encourage better reporting. This process is ongoing and special focus will be placed on those organisations that fail to adequately report.

2.141 Action Point 13 of RHSS concerns public bodies – including government departments and the Civil Service. It requires all public bodies to summarise their health and safety performance and plans in their Annual Reports. This initiative has now been incorporated into departmental mainstream work programmes.

Zurich Strategic Risk/Glasgow Caledonian University Project

2.142 A recent article (July 2004) in CCH's *Business Risk Management Briefing* reported on the Zurich Strategic Risk/Glasgow Caledonian University Project. The project examined the most recent Annual Reports of a small, random set of the FTSE – 350 organisations. The quality of the risk (including occupational safety and health) reporting was assessed, taking into account:

- How specific the statements were.
- How far they went in covering forward plans for dealing with risks through controls, risk mitigation and in explaining risk appetite.
- The context of each occurrence of the word 'risk' and its derivatives.

2.143 This was achieved by categorising each occurrence under the following headings: Treasury, Corporate Governance/Assurance, Strategic/Business, Health/Safety/Environmental, Operational, and Merger/Acquisition. (The last four were combined together as 'Other'.)

2.144 Some of the observations generated by the research include:

- There is a wide variation in the quality of risk reporting.
- Reporting of risk factors commonly features an extensive list of risks with little information on what is being done to mitigate them.
- Those companies that do include commentary on risk mitigation generally seem to engender stakeholder confidence.

- There is little explicit mention of risk appetite in the reports.
- One company recognised that performance measures should not expose shareholders to unreasonable risks.
- If companies perceive value in better reporting of risk, then there is a role for the risk professional in the communication to stakeholders.

2.145 A total of 29 Annual Reports were included in the project. The largest number of times 'risk' and its derivatives was mentioned in a report was 649. The breakdown was: Treasury – 463, Corporate Governance/Assurance – 160, Other – 26. The smallest word count was 21, the breakdown being: Treasury – 4, Corporate Governance – 17, Other – 0. A wide variation indeed!

Institution of Occupational Safety and Health (IOSH) Guidance

2.146 In 2002, the IOSH published detailed guidance on including OSH reporting in Annual Reports in a document entitled 'Reporting Performance', www.iosh.co.uk/technical.

2.147 The guidance is aimed at occupational safety and health professionals and others responsible for internal and/or external/public reporting of organisational health and safety performance. Company Annual Reports provide organisations with a mechanism for reporting their risk appetite, their risk profile and their performance in managing all significant risks, including those in health, safety and environment areas. IOSH recommends the inclusion of summary health, safety and environmental performance results – both positive and negative – in all company Annual Reports as an incentive to achieve the goal of continual improvement.

2.148 Three standards of reporting are outlined in the guidance.

Level 1

2.149 Minimal health and safety reports – these should be issued by all organisations. Typically, findings should be compiled by directors/trustees and presented as a section in the Annual Report.

Level 2

2.150 Comprehensive internal health and safety reports – these should be developed as organisations increasingly accept the business case for good health and safety performance rather than viewing it merely as a compliance issue. Environmental reporting should also be included where relevant.

Level 3

2.151   External health and safety reports – these should be issued to all stakeholders by organisations which value their public image and brand reputation and accept that dialogue with their stakeholders is a key component of corporate governance/corporate social responsibility and, indeed, in their medium-to-long-term sustainability. Again, environmental reporting should be included where relevant.

2.152   For each of the above levels/standards outlined, there should be a common basis. The reports should include:

*   Data on annual outcomes such as accidents, ill-health, lost time, incidents, enforcement action (failure data).
*   Analysis of the data against targets set by the organisation in their annual health and safety plan (success data).
*   Indications of the priorities for the upcoming year, highlighting what has been achieved (success) and what more needs to be done to move towards the goal of continual improvement.

2.153   As organisations increase their commitment to the management of OSH and environmental risks within their workplace, it is likely that its reporting will begin at Level 1, then progress to Level 2 and ultimately to Level 3.

2.154   It is also recognised that many organisations in the private and public sectors currently report internally (Level 2) but do not, as yet, cover health and safety performance data in their Annual Report (Level 3). In such circumstances it may well be necessary to persuade the board – possibly via the assistance of non-executive directors – of the merits of at least Level 1 reporting in the first instance.

2.155   This could manifest itself at the end of the first year as a short overview paragraph within the Annual Report highlighting key points from the internal health and safety report – strengths and weaknesses, successes and failures. In Years 2 and 3 there could be a move towards a fuller, inclusive report and thereafter progress towards a basic Level 3 report.

2.156   The ultimate goal is that all organisations should aspire to Level 3 reporting, either as a stand-alone document or as part of a wider corporate governance/corporate social responsibility report. Indeed, some organisations even produce separate documents for OSH and environmental reporting.

Level 1 – minimal health and safety reports

2.157   Data for these Annual Reports should be compiled from all work activities, including employees, contractors/visitors and members of the public. The following is the minimum that should be included:

- Workplace injuries/ill-health, subdivided by severity into fatalities, major injuries and lost-time events. It is best to present the data as frequency rates in order to allow comparison with previous performance and current/future targets. In addition, the use of such data enables the organisation to benchmark its performance with other organisations.
- In addition to reporting accident and ill-health events it is considered best practice to report total workdays lost per 100 employees/contractors as this provides a measure of both the severity of the injury/ill-health and also of the effectiveness of any associated rehabilitation.
- It is also considered best practice to aim for continual improvement by measuring against national targets such as those set by the HSE in *Revitalising Health and Safety* (RHSS) and also *Securing Health Together*.
- All other significant health and safety events – both positive and negative – should be included, such as: awards won, extended accident-free periods, new initiatives taken and implemented, enforcement action – notices/prosecutions/fines, insurance claims settled.
- Indications of the priorities for health and safety management improvement/action plans and performance targets for the upcoming year

Level 2 – comprehensive internal health and safety reports

2.158 Level 2 reports should include statistical results as well as other performance indicators. Analysis of relevant trends, together with a commentary covering the changes in health and safety performance, should be presented. It is likely that the format of internal reports may initially be different from those aimed at external stakeholders. However, in the interest of openness and transparency, a gradual transition to Level 3 type reporting is recommended.

2.159 Organisations committed to achieving high standards of health and safety performance have found that mere numerical reporting of significant failures and comparison with agreed, set targets (as per Level 1) does not provide sufficient data on which to drive forward improvements. The additional data required may include:

- Systematic causal analysis of injuries, ill-health and damage accidents.
- Causal analysis of major health and safety incidents.
- Commentary on key results of internal audits and inspections.
- External verification/audit report results (eg via insurance company reports).
- Lagging indicators – after the event data – the Level 1 statistical summary of health and safety results – the failure data.

- Leading indicators – positive assurance that best practice aimed at injury/disease prevention are working in practice – before the event/success data.
- Trend analysis of these lagging and leading indicators, using for example a balanced scorecard of different indicators in order to summarise the results.

2.160   Other data that should gradually be incorporated in Level 2 – and then Level 3 – reports include:

- Evidence of management commitment.
- Average health and safety training days per employee.
- Evidence of worker involvement/participation.
- Measurements of health and safety culture – eg climate studies.
- Percentages of risk assessments completed/reviewed.
- Health and safety audit/inspections completed versus target.
- Percentage of completed actions from audits/inspections.
- Percentage of safety-critical maintenance completed on schedule.
- Emergency response exercises held.
- Detailed ill-health data:
    - Number of new cases.
    - Number of days lost.
    - Number of ill-health retirements.
- Insured/uninsured cost of accidents/ill-health.
- Analysis of near-miss events having injury/illness/damage potential.

2.161   In addition to the above trends/indicators, consideration should also be given to the following as health and safety commentary topics:

- Health and safety policy, organisation, responsibilities and accountabilities, arrangements, competent persons, and provision of health and safety advice.
- Occupational health and safety hazards, risk assessment provision, and implementation of suitable and effective control measures.
- The standard of the health and safety management system.
- Internal/external health and safety audit protocols.
- Rehabilitation programmes.
- Employee consultation and participation.
- Training and awareness programmes.
- Management of contractor/supplier health and safety performance.
- Occupational road/travel risks.
- Activities involving external stakeholders.

Level 3 – external health and safety reports

2.162 All the indicators, trends and commentary referred to above for Level 2 should be incorporated into Level 3 (External) reports which should be planned within the context of overall corporate social responsibility (CSR) reporting.

2.163 The Global Reporting Initiative (GRI, 2000) has issued guidelines for public CSR reporting which includes health and safety as one of more than 120 suggested subject areas. CSR reporting processes typically include verification of all internally generated data, thereby increasing their external credibility.

2.164 Level 3 reports should therefore include an assurance that health and safety risks are appropriately included in governance processes in accordance with any national code on organisational risk management.

2.165 Key areas affecting health and safety in best practice external CSR reporting include demonstrating compliance with recognised standards for:

- Board-level responsibilities.
- Assurance/verification processes.
- Good neighbour projects – contractors/suppliers/partners.
- Compliance with International Labour Office codes including:
  - Occupational health management.
  - Human rights.
  - Child labour.
- Implementation of joint health and safety consultation/committees.
- Training for all employees.
- Customer/consumer health and safety monitoring/assurance.

2.166 In addition to commentary on standards compliance, other areas for comment should include:

- Commitment to high health and safety standards.
- Short- and medium-term improvement targets.
- Use of formal management systems.
- Full employee involvement and commitment.
- Recognition of any notable achievements.
- Business impacts of any major accidents.

CONCLUSION

2.167 This chapter has examined the role that OSH professionals can play in the areas of corporate governance (CG), corporate social responsibility (CSR) and business risk management (BRM).

2.168   As organisations get to grips with the increasing need to encompass all three relatively new business requirements, the OSH professional will have an increasingly important role to play at board level in order to ensure that organisations fully adopt BRM/CSR/CG principles and processes. In so doing, organisations will find themselves moving towards the ultimate goal of continual improvement in all their business performance indicators, including continual improvements in their OSH management systems (see **Chapter 3**).

CHAPTER 3

Occupational Safety and Health management systems

INTRODUCTION

3.1 This chapter follows on from **Chapter 2**, which concluded with a reference to occupational safety and health (OSH) management systems. Such systems are vital in terms of ensuring compliance with the requirements of corporate governance (CG), business risk management (BRM) and corporate social responsibility (CSR) principles, concepts, processes and best practices. As with any management system, it needs to be fully integrated into the normal organisational and operational processes of the organisation – built in, not bolted on!

3.2 An overview of existing management systems is included, plus a description of a similar framework – Plan, Do, Check and Act. We refer to quality systems and explain the parallels with effective OSH management systems and the changes to BS 8800. We have described the benefits of organisations integrating their management systems, rather than separate systems and discussed the auditing/verification process.

3.3 As was explained in **Chapter 2**, it is imperative that the OSH professional understands and talks the language of the boardroom so that an OSH risk management system is accepted as part of normal organisational and operational processes. This ensures that OSH considerations are taken into account on a cost versus risk basis, so that the business case for OSH risk management is made using both sides of the cost versus risk equation. On one side of the equation are the costs – which may or may not be losses – and on the other side are the profits (including cost savings) that emanate from effective, proactive business risk management – ie commensurate risk control systems.

3.4 Within the OSH arena, there are legislative requirements to take into account. These force organisations to demonstrate – beyond reasonable doubt (statute law) – that they have taken into account both sides of the equation in terms of 'so far as is reasonably practicable (SFARP)'. As stated in **Chapter 2**, only when the cost is grossly disproportionate to the risk is it not reasonably practicable to ensure that suitable and sufficient risk control systems – as part of an overall OSH management system – are in place and are operational.

3.5   It is therefore important for OSH professionals to be able to quantify both the cost of loss (ie the risk actually resulting in a loss) and the cost of risk prevention (ie the control measures) in economic terms, rather than just stating 'we have to comply with what the law says', which is inevitably a poor motivator, especially at board level.

3.6   This is largely because most boards are remote from the action at the operational levels of their organisations and, although they are considered to be the controlling mind of the organisation, they are – especially in larger organisations – also remote from the risk of individual prosecution under OSH legislation, as cases (including some discussed in **Chapter 2**) have illustrated. Hence there is a push from pressure groups and many sectors of the general public to implement some form of corporate manslaughter/reckless killing legislation that enables boards to be directly responsible for violations of OSH regulations that lead to direct prosecution for corporate manslaughter. The 'controlling mind' of smaller organisations is more easily identified, hence successful prosecutions of directors of smaller companies. However, directors of larger companies regard themselves as immune from prosecution until the law changes. When that happens, we anticipate that boardroom attitudes may rapidly change.

3.7   Having said that, the Health and Safety at Work etc Act 1974, s 37 – highlighted in **Chapter 2** – is still in force but has only been used to prosecute directors and senior managers in small organisations where the linkage evidence is almost self-evident.

3.8   Another approach to be avoided by the OSH professional is one where safety and health is paramount – safety first! The temptation is to get the board to throw money at an issue in the hope that this will resolve the matter. This approach is the result of reactive thinking, is a waste of scarce financial resource, and gives rise to accusations by those outside the OSH profession that OSH professionals are 'risk averse', rather than 'risk managing'.

3.9   The systems approach is by far a better, proactive way of managing all business risks, including OSH risks, and we start this chapter with an explanation of 'management systems' and the benefits in managing OSH risks by using a systematic approach linked to the needs of the organisation and its business and operational processes.

MANAGEMENT SYSTEMS

3.10 Currently there are several comparable management systems used within 'UK plc':

- Quality Management – BS EN ISO 9002: 2000.
- Environmental Management – BS EN ISO 14001: 1996.
- Safety and Health Management – BS 8800: 2004.

3.11 All of these systems follow a similar framework:

- Plan
- Do
- Check
- Act

and all have 'continual improvement' as the ultimate goal.

3.12 The *plan* involves the creation of policies (see **Chapter 4**) that give an overview of what the organisational approach to OSH management is. Part of the plan will involve the organisational framework – the management structure – to ensure that the policy gets implemented. The plan itself – usually an annual plan – should be geared to the goal of continual improvement. It should contain finite goals, targets, responsibilities and accountabilities that are firmly fixed – bound in time – on an individual and collective management basis. Proactive action plans should be an important part of the annual OSH plan and should include aspects such as hazard identification, risk assessment and change management.

3.13 The *do* concerns the actual, practical, day-to-day implementation and operation of the management processes and plans. This is the difficult bit, ie getting the plan to work in practice. Hence worker consultation is a crucial aspect of effective planning and doing. This requires involvement and participation at all levels within the organisation in order to get buy-in and ownership of the overall plan and its component parts. In essence, everyone within the organisation needs to be pulling in the same direction and singing from the same hymn sheet in order to ensure that the plan comes together, is successfully implemented and delivers the ultimate goal of continual improvement in the management system.

3.14 The *check* aspect hinges on the phrase 'What gets measured gets done' and is linked to the target-setting aspects of the plan. In order to motivate individuals and groups to achieve targets aimed at improving the overall system, then accountabilities and timescales for target completion need to be clearly fixed and closely monitored. If a target is set without an end date, it will get put in the pending tray and forgotten about, unless something goes wrong, thereby elevating it up the priority ladder. Similarly, if a target is set and no person is given the accountability for achieving the target, then everyone knows that somebody is meant to be doing something but, in practice, no one does anything! Hence, in order to achieve continual improvement in any management system, SMARTT targets should be set and properly resourced in order to ensure their satisfactory and timely completion. SMARTT stands for:

- Specific.
- Measurable.
- Achievable/agreed.
- Realistic.
- Time-bound.
- Trackable.

3.15   Checking may well involve a combination of proactive and reactive measures, including exposure monitoring, health surveillance, attendance/absence monitoring and the testing of emergency arrangements. Regular checks of any system are vital to ensure continual improvement, as long as any findings resulting from such checks are promptly acted upon!

3.16   The *act* is precisely that – to take action to rectify any concerns flagged up by the checking/monitoring process in order to minimise risk and improve the system. The setting of individual and collective SMARTT targets as described above is imperative to ensure that the right action is taken within the risk timescale. The provision of adequate resources – time, money and competent people – to ensure successful target completion is also a pre-requisite, as are the formulation, development and implementation of individual and collective action plans.

3.17   These four key elements of management systems – PDCA – may be represented as a loop:

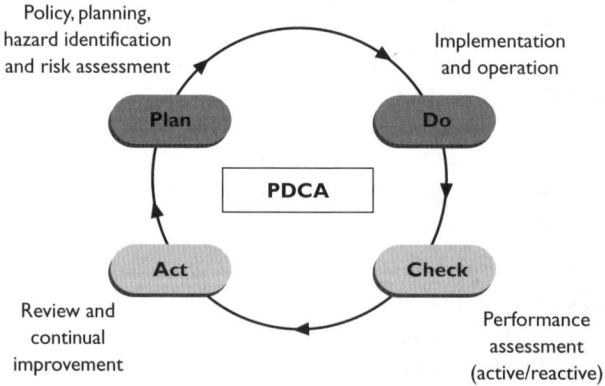

Risk management systems

3.18 The management systems referred to above:

- QU ality.
- EN vironment.
- S afety.
- H ealth.

are all risk-management systems and this integrated systems approach – QUENSH – provides a highly effective route to demonstrating Turnbull/Corporate Governance compliance.

3.19 Established quality assurance procedures provided a sound basis for the development of the majority of current risk management systems. Indeed, the introduction to the Health and Safety Executive's (HSE) first edition of the publication *Successful Health and Safety Management* (HSG 65) in 1991 clearly states that many of the features of effective health and safety management are indistinguishable from the sound management practices advocated by proponents of quality and business excellence. Reference is explicitly made to BS 5750: 1987 – Quality Systems (BSI, London) which has now become BS EN ISO 9001: 2000 – Quality Systems (BSI, London).

3.20 This logic has been extended into the realms of environmental management systems (EMS) initially via BS 7750: 1994 – Specification for environmental management systems (BSI, London), and more recently via BS EN ISO 14001: 1996 – Environmental management systems – Specification with guidance for use (BSI, London), and also into OSH management systems via BS 8800: 1996 – Guide to occupational health and safety management systems (BSI, London). This has recently being updated, with the 2004 version published towards the end of 2004.

3.21 In the original BS 8800: 1996, Annex A clearly demonstrates the commonality of approach between BS 8800: 1996 and BS EN ISO 9001: 1994 – Quality management systems (now superseded by BS EN ISO 9001: 2000).

3.22 Also, BS 8800: 1996 postulates two routes towards successful OSH management systems:

- Route 1 – via HS G 65 (HSE publication).
- Route 2 – via BS EN ISO 14001: 1996 (Environmental management system).

3.23 As BS EN ISO 9001: 2000 (Quality) and BS EN ISO 14001: 1996 (Environment) are both accreditable standards (accredited via approved agencies) and BS 8800: 2004 (OSH) is merely a BSI guide and therefore non-accreditable, BSI – together with some of the accrediting agencies – have developed OHSAS 18001: 1999 – Occupational health and safety management systems – specification. This commercially accreditable standard is not, however, an ISO (International Standards Organisation) approved Standard, unlike the Quality and Environment ISO Standards referred to above.

3.24   More recently (2001), the International Labour Office (ILO) produced Guidelines on occupational safety and health management systems, ILO-OSH, 2001 (ILO, Geneva). These guidelines should be used at national (country) level for the establishment of a national OSH management-system framework, preferably supported by national laws and regulations, and for the development of voluntary arrangements to strengthen compliance with regulations and standards. They should also provide guidance on the development of national and tailored guidelines to respond appropriately to the real needs of organisations according to their size and the nature of their activities. ILO, in developing their guidelines, clearly acknowledges the development of a number of OSH management standards at international, national, regional and industry levels. Whilst these standards are usually well designed and useful, they are not – in ILO's opinion – rooted in the body of internationally agreed OSH principles such as those defined by the tripartite constituency of the ILO. Indeed, only such a linkage – according to the ILO – can provide the strength, flexibility and appropriate basis for the development of a sustainable OSH culture in organisations.

3.25   A long-term implementation of good OSH and environmental practices at all levels of society – ie the continued 'implementation' (compare with Turnbull's 'embedded') of a safety culture as an essential part of general social culture (CSR) – is the only way to curb the spiralling cost of overall health care delivery and environmental protection and remediation while increasing general productivity, according to the ILO.

3.26   As first mentioned in **Chapter 1**, within the broader risk management arena certain national standards have been published such as the Australian/New Zealand standard (Risk Management, AS/NZS 4360, 1996; recently updated in 2004, Standards Association of Australia, Strathfield, New South Wales, www.sai-global.com). More recently, three UK organisations, the Institute of Risk Management (IRM), the Association of Insurance and Risk Managers in Industry and Commerce (AIRMIC) and the Association of Local Authority Risk Managers (ALARM), have collaborated to produce a Risk Management Standard (2002).

3.27   This range of standards will now be examined in more detail.

Quality systems

3.28   The original British Standard on quality systems – BS 5750: 1987 – outlines what was then required by manufacturers and suppliers in order to work to a quality-oriented system. It identified basic disciplines and specified procedures and criteria needed to ensure customer requirements. Within the context of BS EN ISO 9001: 1994/2000, quality means that the product is fit for the purpose for which it has been purchased and has been designed and constructed to satisfy the customer's needs.

3.29   The quality standard sets out how an effective and economic quality system can be established, documented and maintained. It considers that an effective quality system should comprise:

- Management responsibility.
- Quality system principles.
- Quality system audits.
- Quality versus cost considerations.
- Raw material quality control.
- Inspection and testing.
- Control of non-conforming products (failures/rejects).
- Handling, storage, packaging and delivery.
- After-sales service.
- Quality documentation and records.
- Personnel training.
- Product safety/liability.
- Statistical data and analysis.

Quality and safety

3.30   Although the quality standard does not explicitly refer to people safety, there are obvious parallels to be drawn between the quality systems approach and OSH management, as stated in the Foreword to the first edition of the HSE's publication, HSG 65 (mentioned above). Indeed, the management systems described in BS EN ISO 9001 are as applicable to OSH management as they are to product risk management. ISO 9001 is concerned with the achievement of quality, which is measurable against certain specific criteria. It lays down systems that demonstrate achievement against these specified criteria. One of the benefits of an effective quality system is to minimise the risk of product liability claims and losses. In the case of product liability risk management, the specified criteria of performance include:

- Compliance with the relevant consumer safety and consumer protection legislation – eg in the UK, this would include the Health and Safety at Work etc Act 1974, s 6 (as modified by the Consumer Protection Act 1987).
- Compliance with other relevant statutory provisions such as the Management of Health and Safety at Work Regulations 1999 and the Provision and Use of Work Equipment Regulations 1998.
- The ability to adhere to all product contract conditions.
- The minimisation of defective products.

● The maximisation of health and safety benefits to the consumer.

3.31 This parallels very closely with the perceived criteria of an effective OSH management system, namely:

● Compliance with all relevant OSH legislation, codes of practice and standards.
● The ability to adhere to the common law duty of care and relevant aspects of employment contract conditions.
● The minimisation of risks likely to cause injury, disease, damage and/or death.
● The maximisation of OSH benefits to employees, third parties – contractors/visitors/suppliers – and the general public.

3.32 From the above it may be seen that the application of quality systems to OSH management has distinct benefits, especially when consideration is given to the tremendous overlap between the two subject areas. Overlap examples include:

● Policies.
● Systems and procedures.
● Standards.
● Documentation – manuals and record-keeping.
● Training programmes (including training records).
● Statistical analyses – causal, numerical.
● Accident/complaint/defect investigations.
● Audits/inspections (internal and external).
● Taking remedial/control actions.

3.33 Hence, the basis of the vast majority of management systems in current use can be traced back to the quality system framework. In addition, effective quality management systems in the workplace will greatly enhance OSH management practices, thereby leading to an overall and hopefully continual improvement in OSH performance.

Environmental management systems

3.34 The environmental management system standards – initially BS 7750: 1994 and currently BS EN ISO 14001: 1996 – share common management principles with the quality management standard – BS EN ISO 9001: 2000.

3.35 The elements contained in ISO 14001 comprise:

● Initial status review.
● Policy.
● Planning.
● Implementation and operation.
● Checking and corrective action.

- Management review.
- Continual improvement.

3.36 This is usually depicted as a closed loop system – the more times you go round the loop, the better the improvement in environmental performance will be:

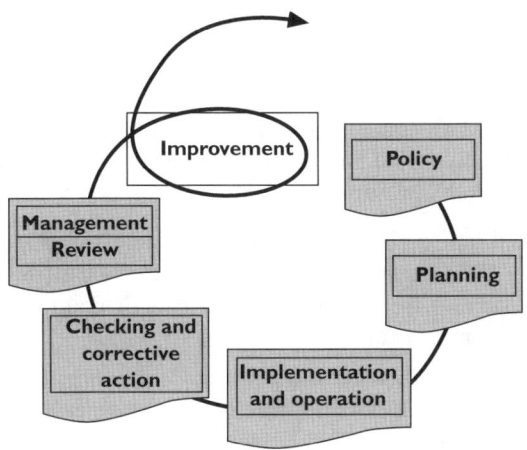

OSH management systems

3.37 The original OSH management system was first depicted in the HSE publication *Successful Health and Safety Management* (HSG 65 1991).

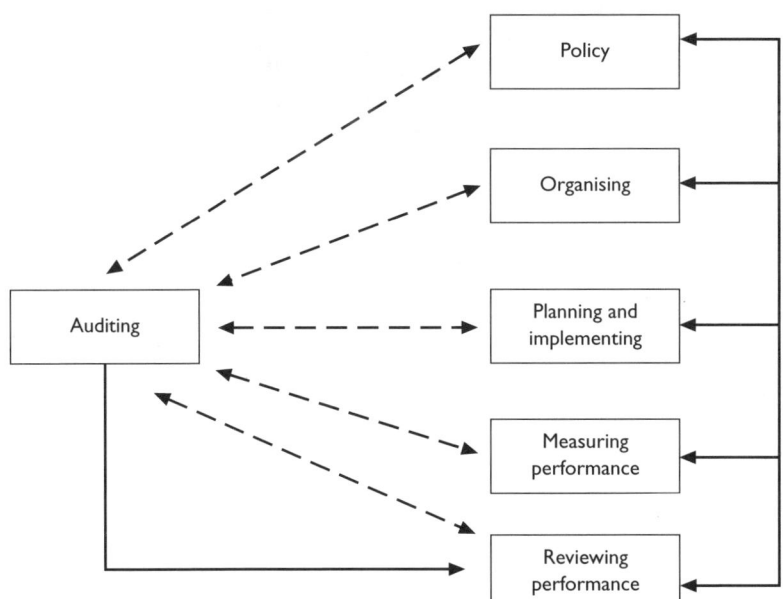

3.38   This led to the mnemonic: POPIMAR:

P	olicy.
O	rganisation.
P	lanning.
I	mplementation.
M	onitoring.
A	udit.
R	eview.

3.39   This POPIMAR framework is the basis of the key stages in the development of an effective OSH management system and policy and is examined in much greater depth in **Chapter 4** of this publication.

3.40   As stated above, BS 8800: 1996 postulated two approaches to the development of OSH management systems: either via the HSG 65 route (see diagram above) or via the ISO 14001 route (see diagram in Environmental Management System, above).

3.41   OHSAS 18001: 1999 also closely mirrors the ISO 14001 approach:

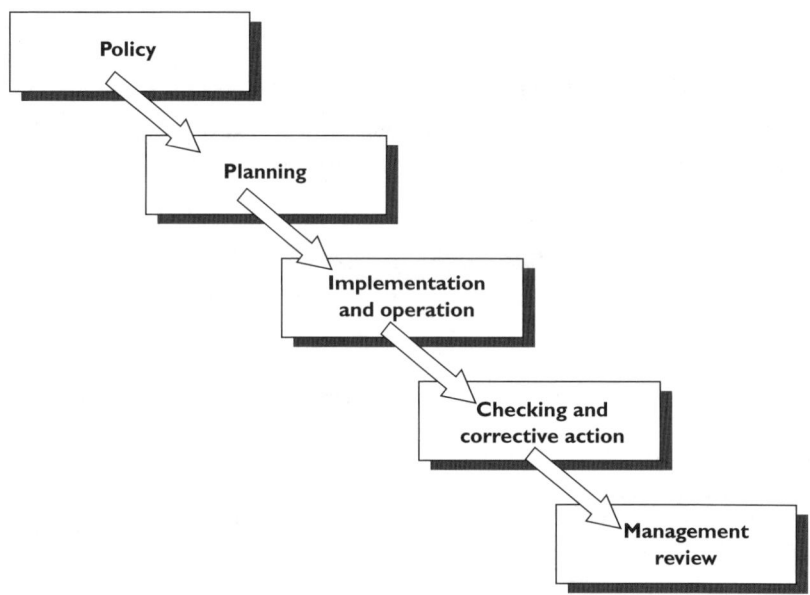

3.42 The ILO Guidelines (2001) approach stresses the need for auditing of all component parts (as does HSG 65) and highlights the need for continual improvement (as does ISO 14001):

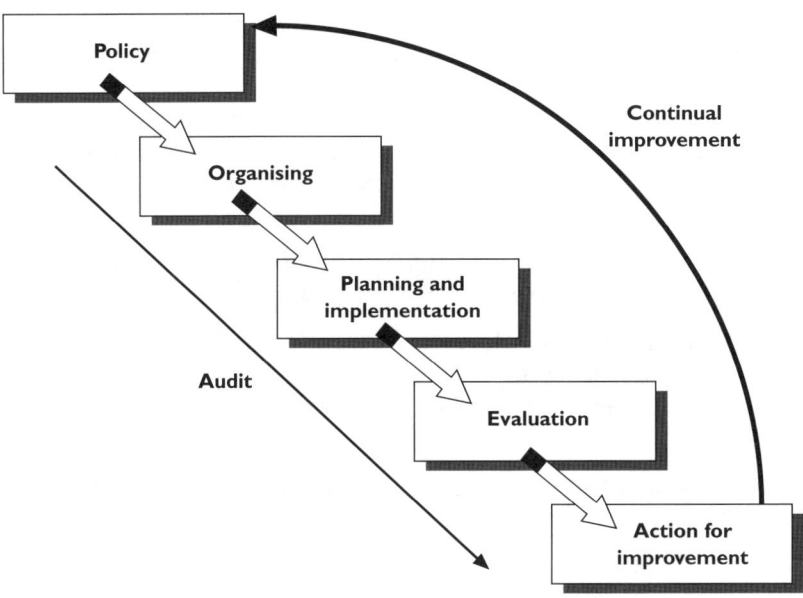

3.43 The following is a useful comparison table showing the various OSH standards discussed above.

Features	Management systems				
	HSG 65	**BS 8800**	**OHSAS 18001**	**ILO**	**Industry-specific (eg responsible care)**
Certifiable	✗	✗	✓	✗	✓
International	✗	✗	✓	✓	✗
Regulator support	✓	✓	✗	✓	✓
Tested	✓	✓	✓	✗	✓
Stakeholder recognition	✓	✓	✓	✓	✓
In-house factors (eg used by customers)	✗	✗	✗	✗	✓

3.44   The revised BS 8800: 2004 (published towards the end of 2004) is still only a Guide. It is not an ISO Standard, nor is it a BS Code of Practice.

3.45   The new version has incorporated the following changes:

- It reflects post-1996 emerging national/international issues.
- It takes into account OHSAS 18001 and, more especially, ILO-OHS 2001 guidelines.
- It concentrates on the HSG 65 and ILO-OHS 2001 approaches.
- It expands the need for an initial status review (baseline audit).
- It stresses the need for continual improvement.
- It recommends the integration of OSH into the organisation's overall management system.

3.46   The aims of the revised BS 8800: 2004 are to:

- Minimise risk to employees and others.
- Improve business performance.
- Assist organisations to establish a responsible image within the marketplace (ie good CSR/CG).

3.47   It is consistent with a number of documents previously discussed in this publication, including:

- HSE's HSG 65: *Successful Health and Safety Management.*
- HSC's ACoP: *Management of Health and Safety at Work Regulations 1999.*
- OHSAS 18001.
- ILO-OHS 2001.
- HSC/HSE OSH management guidance.

3.48   The OSH management system elements or stages are as follows:

- Initial status review.
- OSH policy.
- Organising.
- Planning/implementing.
- Performance measurement.
- Investigation and response.
- Audit.
- Reviewing performance.

3.49   This can be depicted as a flow diagram:

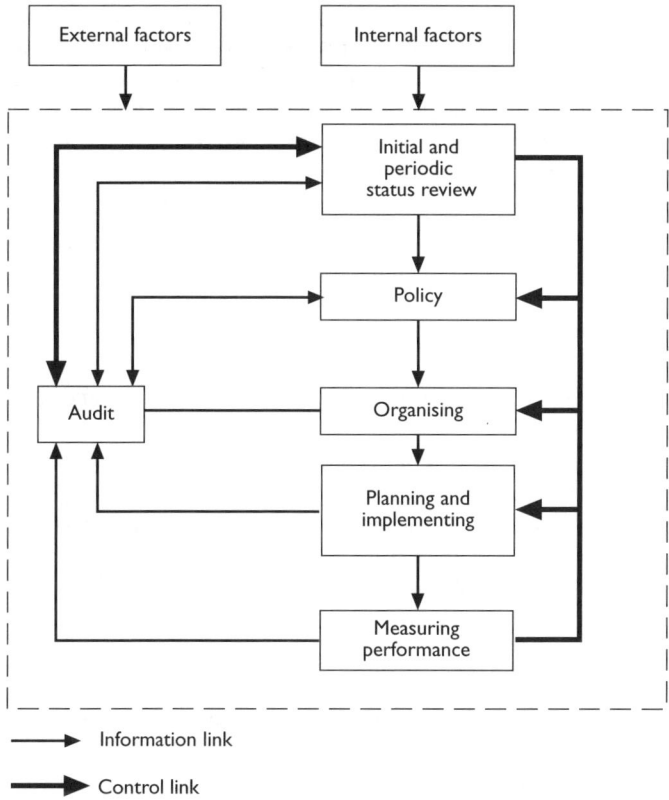

3.50 The Annexes to BS 8800: 2004 are most useful for those wishing to set up or improve an OSH management system. They include:

- Integration with other management system standards.
- Guidance on organising.
- Promoting effective OSH management systems.
- Planning and implementing.
- Risk assessment and control.
- Measuring performance and audit.
- Hazardous event investigation.

3.51 Further information on the content of OSH policies and management systems is contained in **Chapter 4** of this publication.

Advantages and disadvantages of OSH management systems

3.52 The advantages to an organisation of using an OSH management system are many and varied. They include:

- Meeting the risk needs of the organisation by providing a mechanism that allows for prioritisation of risks and commensurate resource allocation to combat them in an efficient and effective way.
- Ensuring that health risks – which are generally less well understood and therefore managed within organisations – are given correct priority.
- Ensuring that OSH risks are given the same importance as other business risks by taking into account both sides of the 'cost versus risk' equation.
- Stressing the need for the QUENSH approach – ie quality, environment and OSH – as one integrated system.
- Easier to attain and demonstrate legal compliance.
- Proving the 'so far as is reasonably practicable' case, as outlined above.
- The ability to achieve the ultimate goal of continual improvement.
- Helping to manage sustainability of OSH initiatives and, indeed, the system itself.
- Demonstrating the visible commitment of directors and senior managers to OSH management.
- A commitment to regular internal and external OSH audits.
- Helping towards corporate governance.
- Reassurance for enforcing authorities and insurers.
- A measured approach to new ideas, thus helping to avoid 'initiative fatigue'.
- Providing a focus for the allocation of scarce OSH resources by ensuring that all parts of the organisation are considered in an objective and equitable manner.
- Ensuring that suitable resources are made available for emergency preparedness, which is part of the OSH management system and should include business continuity considerations.
- Enables defect/non-conformance reporting, which allows operational management to have their finger on the organisation's pulse thereby facilitating prompt, remedial action to be taken so as to minimise the impact of such defects. As stated below, this 'defect in the systems and procedures' reporting is a legal requirement under the Management of Health and Safety at Work Regulations 1999 – reg 14: employees' duties.
- Creating a structural approach for the systematic management of risk, thereby enabling legislative and best practice compliance.

3.53   The disadvantages of using an OSH management system include:

- Too bureaucratic – too much paperwork.
- Integration of management systems may be hampered by internal politics and/or interdepartmental/professional rivalries.
- Too long an implementation time/payback period – seen to be time-wasting in certain quarters, especially before meaningful OSH improvements manifest themselves.
- A heavy demand on sometimes scarce resources – time, money,

people – especially during the initial set-up of the system that may well swamp other equally-beneficial initiatives.

- The fact that aspects of behavioural-based OSH management are not always incorporated in some systems, in spite of the contribution that human-factor considerations, such as monitoring workplace behaviour, talking with and involving people, and minimising human error, are known to make to the continual improvement of the OSH management system.

- The fact that the relatively few certification/accreditation bodies are still on a learning curve, as a result of which different auditors have different interpretations of OSH requirements; however, this should improve as external auditors become more experienced.

- The need to prove true independence of external auditors/certificated bodies, especially as they may also have been involved in providing consultancy to the organisation in connection with the initial system set-up and implementation.

- Any barriers to change that may already exist – or are rapidly developed – within the organisation to slow down a formal OSH management system being introduced.

- Poor understanding of the system by line management that results in a lack of commitment and leadership – probably due to poor planning, communication and initiative overload!

- Audit overload because of a lack of clear guidance, planning, explanation and management support – 'You are the fourth group of people that have been round this place in the last six weeks!'

- Choosing the right OSH management system from those outlined below that can be adopted, adapted and improved to the specific local, national and international needs and culture of the organisation; otherwise it will not be sustainable.

- Ensuring that the written procedure/system is, of itself, 'safe and healthy' – ie not just to write down what is actually done in practice, but to ensure that what gets written down is achievable and also ensures total legislative and standard compliance as a minimum so that the system is comprehensive (covering all identified hazards) and adequate (in terms of having a raft of effective controls designed to reduce risk).

The UK risk management standard

3.54   As stated above, the standard was produced in 2002 via a collaboration of three risk management organisations: IRM, AIRMIC and ALARM. The standard was needed in order to ensure that there was an agreed:

- Terminology related to the words used: wherever possible the terminology for risk set out by the International Organisation for Standards

(ISO) document – ISO/IEC Guide 73 Risk Management – Vocabulary – Guidelines for use in standards – has been used.

- Process by which risk management can be carried out.
- Organisation structure for risk management.
- Objective for risk management.

3.55 Importantly, the standard recognises that there are both speculative and pure risks, ie risk has both an upside (gain) and downside (loss).

3.56 The standard also recognises that risk management protects and adds value to an organisation and its stakeholders via support for the organisation's goals and objectives by:

- Providing a framework for an organisation that enables future activity to take place in a consistent and controlled manner (ie Turnbull/ Corporate Governance).
- Improving decision-making, planning and prioritisation by comprehensive and structured understanding of business activity, volatility and project opportunity (gain)/threat (loss).
- Contributing to more efficient use/allocation of capital and resources within the organisation.
- Reducing volatility in the non-essential areas of the business.
- Protecting and enhancing assets and company image (good CG and CSR).
- Developing and supporting people and the organisation's knowledge base.
- Optimising operational efficiency.

3.57 The risk management process outlined in the standard may be depicted as follows:

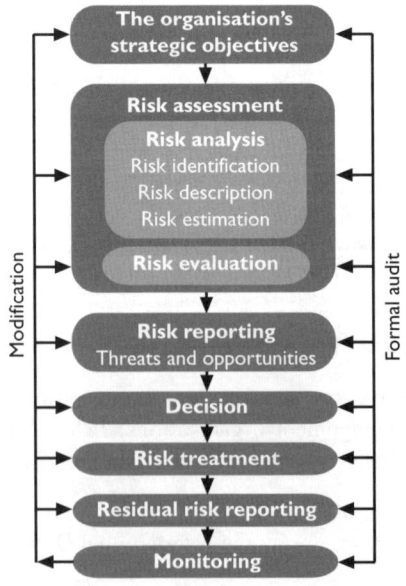

3.58   The section on risk reporting and communication gives guidance on both internal and external reporting.

3.59   In connection with internal reporting, the standard recommends that the board of directors should:

- Know about the most significant risks facing the organisation.
- Know the possible effects on shareholder value of deviations to expected performance ranges.
- Ensure appropriate levels of risk awareness exist throughout the organisation.
- Know how the organisation will manage (in) a crisis.
- Know the importance of stakeholder confidence in the organisation.
- Know how to manage communications with the investment community where applicable.
- Be assured that the risk management process is working effectively and efficiently.
- Publish a clear risk management policy covering risk management philosophy and responsibilities.

3.60   As far as external reporting is concerned, the standard recommends that an organisation needs to report to its stakeholders on a regular basis – at least annually – setting out its risk management policies and the effectiveness in achieving its objectives. Increasingly, stakeholders look to organisations to provide evidence of effective management of its non-financial performance in such areas as community affairs, human rights, employment practices, OSH and the environment.

3.61   Good corporate governance requires that organisations adopt a methodical approach to risk management that:

- Protects stakeholder interest.
- Ensures that the board of directors discharges its duties to direct strategy, build/add value and monitor the performance of the organisation.
- Ensures that management controls are in place and are performing adequately.

3.62   The arrangements for the formal reporting of risks (ie risk profile) and their management should be clearly stated – possibly via an organisational risk management policy – and should be made available as a matter of course to all stakeholders.

3.63   The formal reporting should address:

- The control methods, particularly management responsibilities and accountabilities for the risk management process.
- The processes used to identify and assess risks, and thereafter how they are addressed by the risk management system.

- The primary control systems in place to manage those risks assessed as significant.
- The monitoring and review system in place.

3.64 Any significant deficiencies uncovered by the system – or of the system itself – should be reported to the person/group charged with risk management responsibilities/accountabilities, together with the control measures/remedial actions taken to deal with them.

3.65 The standard clearly defines the role and responsibility of the board as to determine the strategic direction of the organisation, and for creating the environment and the structures for risk management to operate effectively and efficiently. This may be achieved via a delegated executive group, a non-executive committee (to advise the board), an audit committee, or such other function that suits the organisation's way of operating. This group/committee should be capable of acting as the sponsor and focal point for risk management within the organisation and beyond.

3.66 The board should, as a minimum, consider the following in evaluating its system of internal control (ie Turnbull compliance):

- The nature and extent of downside (loss) risks acceptable for the organisation to bear within its particular business.
- The likelihood of such risks becoming a reality.
- How unacceptable/significant risks should be managed.
- The organisation's ability and competence to minimise the probability and impact on the continued running of the business.
- The costs and benefits of the risk and associated control measures – ie the total cost of risk.
- The effectiveness and efficiency of the overall risk management process.
- The risk implications – both downside and upside – of board decisions.

3.67 The full risk management standard is available from the IRM website (www.theIRM.org).

SYSTEMS IN FOCUS

3.68 The Institution of Occupational Safety and Health (IOSH) have produced strategic corporate guidance on the issue of OSH and other management systems entitled *Systems in Focus*. The document, produced in 2002, is available via the IOSH website: (http://www.iosh.co.uk/files/technical/TIG SystemsinFocus0302wv%2Epdf).

3.69 The document examines the following areas:

- The main components of OSH management systems.
- An overview of typical systems.

- Regulatory/industry standards – global perspectives.
- Whether management systems should be integrated.
- Key features of OSH management systems:
  - Continual improvement.
  - System activities.
  - Stakeholder involvement.
  - Auditing/verification.
- Advantages and disadvantages of OSH management systems.
- OSH management certification.
- How to get started.

Main components of OSH management systems

3.70 As stated above, all systems are based on PDCA – Plan, Do, Check, Act.

3.71 Effective OSH management systems include the following elements:

- Policy.
- Planning.
- Organising.
- Role of employee representatives.
- Communicating.
- Consulting.
- Implementing and operating.
- Measuring performance.
- Corrective and preventive actions.
- Management review.
- Continual improvement.

3.72 This structure/framework is similar to that of POPIMAR, referred to above and in more detail in **Chapter 4**. Briefly, the functions of the various elements of the system are as follows.

Policy

3.73 A statement of commitment and vision by the organisation that creates a framework for accountability, incorporated and led by the most senior management

Planning

3.74 Plan for hazard identification, risk assessment and risk control, emergency preparedness and response, with identification of legal and other standards that apply. The organisation should set short, medium and long-term

OSH objectives and plan the management targets and actions needed to achieve them.

Organising

3.75 Defines the organisational/management structure; allocates OSH responsibilities and accountabilities to managers, supervisors/team leaders and employees that should be linked to operational control arrangements. The four Cs – control, co-operation, communication and competence – are vital in this regard (see **Chapter 4**).

Role of employee representatives

3.76 They are a crucial skill resource that will make a valuable contribution to the overall organisational response to the management of risk.

Communicating

3.77 From basic information and work procedures to the details of the system itself, in two-way communications – ie upward and downward. This element is discussed in detail in **Chapter 6**.

Consulting

3.78 Whatever the flow of information, effective mechanisms are required to tap into the fund of knowledge and expertise within the workforce, clients, suppliers and other stakeholders such as regulators, trade unions and neighbours, and to facilitate the collective shaping of the risk management programme.

Implementing and operating

3.79 The implementation and operation of individual and collective management action plans and processes; engaging in the activities from hazard identification and risk assessment through to audit and review – ie the practice of the OSH management system.

Measuring performance

3.80 From reactive data on the frequency and severity rates of work-related injuries, ill-health, diseases, near-misses and other incidents, through to active data on routine inspections, OSH committees, training, risk assessments, etc, including formal audits that evaluate the overall performance of the OSH management system.

Corrective and preventive actions

3.81 This is a fundamental OSH management system component and should comprise a systematic approach to identifying opportunities to prevent accidents and ill-health via their investigation. A variety of techniques should be used to identify non-compliance and thereafter correct them by seeking ways by which adverse outcomes may be prevented. The best lesson to learn from an accident is how to prevent it happening again!

Management review

3.82 This element evaluates the appropriateness, effectiveness and efficiency of the overall system design, its resourcing and its objectives in the light of the actual performance achieved. This enables future goals to be set in order to continually improve over time.

Continual improvement

3.83 The heart of any OSH management system is a fundamental desire and commitment – throughout the organisation – to be better at managing OSH risks proactively, so that accidents and ill-health are reduced (effectiveness) and/or the system achieves the desired aims/goals/targets using fewer resources (efficiency).

Typical systems – overview

3.84 This has been explored above where systems such as HSG 65, BS 8800, OHSAS 18001, and ILO-OSH 2001 were described.

Regulatory/industry standards: global perspective

3.85 To aid the reader in getting a broader, non-UK perspective, we have set out below brief details of other regulatory/industry standards and/or legislation covering aspects of OSH management systems.

- EU management systems/arrangements (eg MHSWR – Management of Health and Safety at Work Regulations 1999):
  - Risk assessment: key processes (eg MHSWR).
  - Safety case regime for major hazards (eg COMAH – Control of Major Accident Hazards Regulations 1999).

- Canada management systems for OSH:
  - Due diligence.
- Norway/Sweden – OSH internal control systems (mandatory).
- India (post-Bhopal) – systematic disaster-prevention management.
- China – has adopted ILO-OSH 2001 guidelines.
- Australia/New Zealand – have well-developed national standards for OSH management. ANSI Z10 OSH Management Systems standard imminent (2004):
  - Safety case regime for major hazards.
- Pacific Rim – have adopted OSH management systems
  - Have third-party/external auditing (government approved)
- USA – Voluntary Protection Programme (VPP) leading to exemption from OSHA inspection.
- USA – ANSI ASC Z10 Occupational Safety and Health Systems.

Integration of management systems

3.86 The integrated management systems (IMS) approach is the preferred option for many organisations. The QUENSH approach has been referred to above, where quality, environment and OSH systems have become fully integrated and the plan-do-check-act loop is circumnavigated; not three separate times for each discipline, but only once for all three.

3.87 The benefits are:

- Better efficiency.
- More effective as subject areas are not treated in isolation – ie the silo effect is minimised and the gaps are minded.
- Better Turnbull compliance.
- Best practice in business risk management (BRM).
- Support to brand reputation and company image.
- Support to business continuity.
- Minimisation of negative financial impact.
- Better overall risk-profiling and hence prioritisation.
- Better allocation of resources – time, money, people.
- The interaction between different risk areas is taken into account.

3.88 Integrated management systems need to be developed, implemented and monitored by persons competent in the overall business risk management (BRM) process and the particular subject area – for example, OSH – under review. The lay definition of competent person is someone who knows what to look for, knows when they have found it and knows what to do about it, but also knows and reports when their competency in a particular area of activity is not sufficient!

3.89 The ultimate goal of all IMSs is that of continual improvement. This drive for continual improvement in all BRM areas can be further enhanced by the setting of SMARTT targets, establishing agreed key performance indicators (KPIs) and the use of regular (at least quarterly) performance appraisals to fix accountabilities for all directors, managers and supervisors who contribute to the achievement of the organisation's goals, vision and mission. An effective IMS greatly enhances OSH management and leads to continual improvement in the level of performance.

Key features of OSH management systems

3.90   There are four key features that should figure in all OSH management systems:

1   Continual improvement.
2   System activities.
3   Stakeholder involvement.
4   Auditing/verification.

Continual improvement

3.91   Continual improvement is the ultimate goal, ie better year-on-year results in terms of reducing numbers of injuries, diseases, damage accidents and death. This should manifest itself in a steady improvement in results using fewer resources over time. This shows that the OSH management system is both efficient and effective as a result of better targeting. Eventually, if all goes according to plan – there has to be a plan – then the watershed of breakthrough performance is reached when the results move the OSH culture of the organisation to a new enhanced state. 'Zero accidents' really means zero accidents! The system has become better, more comprehensive, easier to understand and accepted throughout the organisation. But – woe betide – do not let complacency lead to a false sense of security because once organisations revert to type and start going down the slippery slope of poor practice in OSH management, all too rapidly they reach the point of no return – a disaster or catastrophe like those discussed in **Chapter 2**! It is a very long way back up the slope, and some organisations never make it!

3.92   Continual improvement may be achieved via a consistent and objective approach to OSH management, utilising a range of skills and practices known to the OSH professional:

●   Regular OSH audits, both internal and external.
●   Statistical improvements – target-setting.
●   Benchmarking with others in similar industries/sectors.
●   Industry/sector guidelines on best practice.
●   Ownership of work processes by managers, supervisors and safety representatives – team-working (discussed in **Chapter 6**).
●   Active monitoring (not just reactive, negative statistics).

- Workforce involvement, which greatly assists the efficient and effective operation of the system.
- Diagonal slice (through the organisation's structure) improvement/task groups.

3.93 The use of improvement/task groups for OSH problem-solving is an excellent way of capturing ideas for making the system better. They enable full and frank discussion of ideas and feedback to all concerned on the generated solutions. Via such teamwork, ownership and buy-in are rapidly secured, as opposed to the more traditional way of imposing the will of management onto employees via senior management presentations or, even worse, a memo pinned to the OSH notice board!

3.94 The end product of any improvement group should be an agreed action plan with clearly allocated responsibilities and timescales so that everyone knows who is going to do it, and by when. All plans need to be properly resourced, implemented and monitored. It is vital that senior management give visible and continuous support and commitment to these OSH action plans. The use of SMARTT targets (as described above) will greatly assist in this regard.

System activities

3.95 System activities should be well documented and available on a 24/7 basis in both paper and electronic format throughout the organisation. An increasing number of organisations have moved away from paper-based voluminous OSH manuals and procedures to having an OSH section on their Intranet (internal company 'Internet'). As long as the worker on the Sunday night shift has access to the information stored electronically – ie has a password and knows how to use the system – then this is acceptable. Where any doubt regarding access/availability exists, then paper back-up should be made available.

3.96 In addition to the provision of the OSH management system manual – either paper or electronic versions – there is also a need to train all employees who have been given OSH responsibilities and accountabilities in order to ensure that the agreed SMARTT targets are resourced and achieved. A clear understanding of the standard to be met, together with the audit requirements, needs to be thoroughly explained via such training sessions, all of which should be recorded.

3.97 OSH management systems should have clearly-defined objectives and targets to be achieved during the ensuing year. This setting of objectives greatly assists in deciding what practical activities are needed to meet the system requirements. Such high-level objectives may include:

- Clear OSH policy.
- Written commitment to the OSH policy by the board/CEO.

- Demonstrable, acceptable standards.
- Visible leadership.
- Adequate resources.
- Personal involvement at all levels of management/supervision.
- Regular proactive and reactive monitoring.
- Performance appraisal reviews.
- Employment of competent, resourced, trained and developed employees.
- Effective stakeholder involvement/consultation.
- Ensuring that all purchases – equipment, materials, services (eg contractors) – are selected using OSH criteria, not just based on price.
- Making sure that all necessary technical and operational data/records are available on site. These should be easily accessible, up to date, regularly reviewed, and retained so as to ensure demonstrable business/regulatory compliance (eg Turnbull, Corporate Governance).
- Regular 'sharp-end' monitoring to ensure that actual performance measures are up to expected performance – mind the gaps!
- Pre-planned audits to verify the practical effectiveness of the OSH management system.
- Prompt reporting of non-conformances – eg accidents, near-misses – with follow-up root-cause analysis and commensurate corrective actions.
- Well-defined and tested emergency systems:
  - Plans.
  - Competent implementers (eg fire wardens).
  - Aims: to contain business failures and to minimise adverse effects.

3.98 As a result of the above 'objectives' list, the type of activities that should be taking place on a regular – almost daily – basis within organisations include the following:

- Hazard management:
  - Are they novel?
  - Are they understood by all concerned?
  - Are internal/external stakeholders at risk?
  - Will the hazards have short or long term effects if not properly managed?
- Organisational structure/type considerations:
  - Is it a single-site location?
  - Is the organisation multi-site and/or multinational?
  - Are there a few/many customers?
  - Where are the key dependencies for the critical business functions – eg outsourcing, single-source supply, just-in-time delivery.
- Consideration of the range of technologies/disciplines needed within the organisation:
  - Many/few standards to be met?

–   Employment of competent persons: square pegs in square holes (compare with the Management of Health and Safety at Work Regulations 1999, reg 13: employees must be physically and mentally capable of doing their job without adverse risk to their – or others – health and safety).

● Compliance with legislative and other applicable standards:
–   OSH, environmental, quality, etc?
–   Prescriptive laws or goal-setting?
–   External operating standards?
–   International dimension/ramifications?

3.99   Each of the OSH management activities should be in the form of an auditable standard – ie a system or process – that has defined inputs to achieve defined outputs, ie the goals. These goals may include:

● OSH policy that is up to date, signed, dated and communicated throughout the organisation, and is regularly reviewed (at least every three years).

● A hazard register that contains all identified hazards, related risk assessments and control measures, together with the consequences (minimum to maximum) of any foreseeable failure.

● The categorisation of reported accidents by potential (not actual) outcome. This involves asking the question: what could have been the worst possible outcome – the maximum potential loss (MPL)? The follow-up investigation's depth should be based on the MPL and should involve root-cause investigation so far as to establish all the contributing causes with suitable, trackable recommendations designed to eliminate these causes in the future. The recommendations – responsibilities, timescales, etc – need to be trackable so as to ensure close-out and sign-off at the end of the investigation.

● OSH performance criteria should be used to award and manage any contracts entered into by the organisation.

Stakeholder involvement

3.100   Stakeholder involvement needs to include individuals and groups, both internal and external to the organisation, all of whom are affected by the operation of the OSH management system and are therefore potentially interested in its content and effectiveness.

3.101   Internal stakeholders include directors, trustees, workforce, trade unions, safety representatives, on-site contractors and OSH professionals.

3.102   External stakeholders include: regulators, neighbours, clients, customers, supply-chain contractors, insurers, shareholders/investors, banks, lobbyists, pressure groups, global organisations (eg ILO).

3.103 One of the most important stakeholders – if not the most important – is the workforce. This is because generally they are the most 'at risk' group. Therefore they need to be consulted on a regular and frequent basis because they have first-hand knowledge of workplace hazards and, hence, their controls:

- Do existing controls work in practice?
- Are they efficient and/or effective?
- Are there better, more cost-effective, safer ways of getting the job done?

3.104 These are the sorts of questions to be asked and answered via joint consultation. Indeed, through worker involvement, the solutions arrived at will more than likely be adhered to because of ownership and buy-in. In some cases the solutions are more dynamic and forceful than any management may wish to have imposed! It has been demonstrated – time and time again – that employees are excellent sources of ideas and suggestions for continual improvements in the areas of OSH, environment and quality (NB QUENSH).

3.105 Hence, it is important to ensure that the entire workforce has been trained in hazard identification and awareness, especially what has become known as the 'mechanism of harm'. If employees understand how noise makes you deaf or how certain chemicals can scar the lungs, they are much more likely to understand and adhere to the commensurate control measures. It is essential to develop this understanding – this workforce memory – especially in connection with medium-term, disease-related effects.

3.106 To this end, it should be remembered that trade unions and worker OSH representatives generally have a wider knowledge of and a strong commitment to OSH and hence are a highly significant – and sometimes under-utilised – resource to be incorporated into the OSH management system. Whatever formal systems and controls are in use, the individual or group performing a task has a great influence over the outcomes. Legally and morally, each person has a duty of care to themselves and others who may be affected by the acts and omissions. Individual and group behaviour is largely governed by peer pressure; hence the need for full hazard awareness and understanding at local level, so as to ensure that the exerted peer pressure is of a positive, reinforcing nature that results in healthy and safe behavioural patterns in both individuals and groups. This is the basis of behavioural-based OSH management systems.

3.107 It is imperative that workforce OSH representatives are seen as key stakeholders, as stated above. Indeed, there is a EU legal requirement for them to be consulted by management on OSH-related matters and concerns. In the UK this requirement has manifested itself as the Health and Safety (Consultation with Employees) Regulations 1996. The reasons for including these key stakeholders in the OSH management system have been briefly mentioned above and are as follows:

- They have detailed knowledge of what actually happens – as opposed to what should happen – at the sharp end.
- They are able to identify improvement opportunities that are realistic and achievable in practice.
- They should be involved in joint workplace inspections, at least on a quarterly basis, as part of an ongoing, proactive monitoring regime.
- They should be involved in the root-cause analysis approach to accident investigations, especially those reported under the Reporting of Injuries, Diseases and Dangerous Occurrences Regulations 1995 (RIDDOR).
- They should be the focus of all employee questions, concerns and reports on 'shortcomings' in the OSH management system (see the requirement under the Management of Health and Safety at Work Regulations 1999, reg 14).
- Because of their linkages with external bodies, such as trade unions, they have access to 'best practice' OSH information.
- They inevitably can act as a 'reality check' on the overall OSH management system.

3.108 Another key stakeholder group are OSH professionals. It is in the interests of all OSH professionals to be advocates of the OSH management system operating within their organisations, and also advocates of the benefits such a system brings to the credibility and continued success of the organisation.

3.109 Under the Management of Health and Safety at Work Regulations 1999, reg 7 the employer must appoint one or more competent persons to assist in undertaking the measures he needed to comply with the requirements (the dos) and the prohibitions (the don'ts) imposed by or under the relevant statutory provisions. Where more than one person is appointed, teamwork is vital to ensure that the OSH management system is comprehensive, efficient and effective.

3.110 OSH professionals have a key role in advising others having OSH management responsibilities and accountabilities, especially in connection with OSH hazards, their likely effects and current best practices for hazard avoidance, minimisation, control and mitigation. For OSH professionals to be able to give competent advice, they need to understand the ramifications of relevant legislation, codes, guidance, standards and best practice. Also, knowledge of practical risk-assessment methods, cost-beneficial control methods and the provision of training and communications programmes are essential. From a 'Professionals in Partnership' viewpoint, the OSH professional is also likely to liaise with internal and external professionals in related disciplines such as occupational health/medicine, occupational hygiene, ergonomics, psychology, environmental management, engineering, construction, etc.

3.111   It is highly likely that the OSH professional will be appointed champion or custodian of the OSH management system on behalf of the organisation, with a key role for effective implementation and regular review. OSH professionals should also be able to make key contributions to audit processes and investigations of serious non-conformances within the system – ie injury, ill-health or damage.

3.112   Key external stakeholders include: regulators, CSR lobbyists and global bodies.

3.113   Regulator actions reflect growing intolerance by society at large for those organisations whose profits appear to be earned without due care and consideration for the health and safety of employees, clients, customers or the general public.

3.114   Increasingly, as outlined above – especially in Pacific Rim countries – it is becoming noticeable that legislation is beginning to require certification to a recognised national or international OSH management standard. Within the UK the requirement for safety cases in high-hazard industries such as offshore and railways ensures that summaries of the OSH management systems are included in the submissions made by the operator to the regulator.

3.115   Investors and insurers are concerned about risk for different reasons but both are essentially looking for well-managed risks. Whereas Turnbull requirements are initially targeted at avoiding major financial and non-financial losses, investors require more positive reassurance that a business is well-managed. They may take OSH performance as a marker of performance on trade type unless they can be convinced – via organisational surveys and discussions – that OSH/risk issues are being well-managed. Evidence of a comprehensive, proactive and effective OSH management system is vital in convincing investors and especially insurers that the organisation is a good quality risk.

3.116   The CSR Lobby is ensuring – especially in the developed countries – that demand is growing for organisations to visibly and publicly demonstrate their CSR credentials. As has been stated in **Chapter 2**, a number of stock markets now operate investor listings linked to CSR results that include OSH considerations. Indeed, a growing number of non-governmental organisations (NGOs), investors and consumers ask questions about OSH performance, especially when suspicions exist that jobs may be exported to locations having less stringent OSH requirements and standards than the parent company's home country. In any event, the parent company should impose best practice onto any overseas operations – that means the same OSH management system being in force, irrespective of geographical location – ie global best practice (see also reference to SA8000 in **Chapter 2**). In addition, there is a growing consensus that the effective management of risks needs to extend throughout each organisation's supply chain so as to ensure security of supply and thus sustainability, a development reflected in the ILO OSH 2001 guidelines discussed above.

3.117 Global bodies – such as the United Nations, International Labour Organisation (ILO) and the World Health Organisation (WHO) – set a variety of OSH standards that are sometimes based on the Universal Declaration of Human Rights. With globalisation, such standards assume higher profiles and responsible organisations should pay more attention to ensuring compliance, particularly in areas such as protection of vulnerable groups:

- Children.
- Migrant workers.
- Female workers.

3.118 Compliance with these standards is often a requirement for operating in developing economies, in particular for projects funded by the World Bank or the International Monetary Fund. Third-party verification of such compliances may also be required, hence the link to OSH management systems.

Auditing/verification

3.119 Auditing may be defined as 'the sampling of a process by a competent person or persons who are independent of that process'.

3.120 Auditors should report on the effectiveness and efficiency of the process via an examination of the system inputs, outputs and internal controls. In essence this is a broadbrush approach with a brief look at all aspects of the system's operation. It is not an in-depth drill-down type of survey. It should highlight both strengths (best practice) and weaknesses (areas where improvements are necessary). We call these 'horizontal' and 'vertical' audits respectively, and they are discussed further in **Chapter 7**.

3.121 Verification, usually via external audit/auditors, leads to the issue of a compliance certificate – certification – as with ISO 9001: 2000 and ISO 14001: 1996, described above.

3.122 Typical audits should comprise three component parts:

1   Documentation review:
     - Are they adequate?
     - Up to date?
     - Available?
     - Reflect all hazards?
     - Known, understood and adhered to?
2   Interviews to confirm awareness, competence, adequacy of resources, understanding.
3   Observation to check that what is written actually happens in prac-

tice – look for the gap(s) between the documentation and actual practice in the workplace!

3.123   Key audit features from a best practice viewpoint include:

- Independence of auditor(s) (of the process or organisation being audited).
- Credibility and competence of auditor(s).
- Analytical ability and people skills (eg interviewing) of auditor(s).
- Sampling of organisation's general and local documentation to gain evidence to demonstrate that operational practice is consistent with what is written.
- Getting further supportive evidence (or otherwise) via additional documentation, interviews and job/task observations – the walk-round survey visit.

3.124   It should be borne in mind, however, that sampling does not give the complete picture (ie the perfect view); it is merely a snapshot in time and is therefore only valid at the time of the audit.

3.125   From a best practice viewpoint the following practical guidance should be taken on board in developing a robust audit protocol:

- Do not allow a 'contest' mentality to develop; league tables based on overall scores may seem useful at board level, but who gets promoted and who gets relegated?
- Ensure that any hidden weaknesses are highlighted in a constructive manner.
- Ensure strengths/successes are equally highlighted – praise where praise is due!
- Use transparent, agreed performance standards.
- Bear in mind that audits are not the only source of ideas for continual improvement (eg suggestion schemes, focus/task groups).
- Accentuate the positives (strengths), list the negatives (weaknesses) for further discussion and recommendations for improvement.
- Ensure that the final audit report clearly states what is expected, by whom and when.
- Ensure the audit plan covers all organisational areas at least on an annual basis.
- Scoring systems should encourage future improvements via the use of benchmarking.
- Consideration should be given to expanding the audit team to include key stakeholder representatives.

3.126   Specifically, as far as competence is concerned, each auditor on the team – ideally a team of three (not with two away!) – needs to have necessary knowledge (of OSH and the processes involved), skill (to ask the right questions at the right time and be able to correctly interpret the answers given), practical experience (of the audit process) and personal qualities (eg people skills). This competence needs to cover two main areas: auditing methods and the processes being audited. The teamwork approach mentioned above allows new auditors to be developed but also allows a consensus to be reached in connection with the more difficult questions/subject areas – three heads are better than two! One measure of acceptable competence is considered to be an OSH qualification at a high level such as NVQ4 in OSH Practice or the NEBOSH National Diploma. Additionally, three years' OSH/auditing experience coupled with some form of regulated continuous professional development (CPD) would be considered acceptable.

3.127   The question of using internal or external auditors needs to be answered. There are advantages and disadvantages on both sides, so the decision is down to the organisation. A useful compromise may well be to mix up the team – ie two internal and one external!

3.128   The advantages of internal auditors are that they:

- Know the organisation and where to look for evidence.
- Have high internal credibility as they are known within the organisation.
- Will produce findings viewed as realistic, because of peer review.
- Achieve excellent developmental experience as internal auditors, which benefits the organisation.
- Share opportunities during the audit process which greatly aids the transfer of best practice throughout the organisation.

3.129   The disadvantages of using internal auditors include:

- Suspicions by external stakeholders that internal auditors are not sufficiently independent.
- Internal auditors take scarce resources away from normal operations.
- Internal auditors may have limited vision concerning improvement opportunities because of their lack of knowledge of external benchmarks/best practice.

3.130   The advantages of external auditors are that:

- They have high credibility amongst external stakeholders.
- They bring strong benchmarking knowledge and experience to the team.
- They can provide access to external verification bodies that may well result in added value via certification/accreditation.

3.131   The disadvantages of using external auditors include:

- The costs versus the benefits – external audits can be more expensive than the cost of internal resources.
- The fact that external auditors have to earn organisational respect

and credibility; initially they will be viewed negatively – they are out to catch us out!

- Not knowing the organisation and the people involved, hence a longer learning curve/timescale, especially as more documentation will need to be provided and digested.

3.132 We have a solution to many of these issues, which is outlined in **Chapter 7**. The tool we use looks at the organisation from a systems perspective and takes a broader review of the organisation, coupled with a review of the OSH management systems and its interrelationships with other business and operational process. The tool also reviews the status of the OSH management system in respect of organisational factors and risk profile. These aspects are also discussed in **Chapters 6 and 7**. As a consequence, the organisation obtains an overview status report, which identifies the key risks and where a detailed audit may be necessary.

3.133 The mix of the audit team is important; although, having said that, only one external auditor, primarily because of the costs involved, undertakes the vast majority of accreditation/certification audits.

OSH management system certification

3.134 The desire for accreditation/certification of the OSH management system by an external body inevitably comes from internal stakeholders, which results in a boardroom decision to get 'certified'. This decision in itself has cost versus risk implications and is usually made once the board have been convinced of the benefits certification brings from a commercial/CG/CSR viewpoint by adding value with clients and consumers.

3.135 In some cases, the pressure to make the decision may well come via external stakeholders, even from non-executive-director level or, more likely, from potential or existing clients, local regulators or companies in the supply chain. This may lead to the fast-track approach to OSH management system development and implementation as certification moves rapidly from being a 'preferred option' to becoming a 'condition of entry' into a new marketplace.

3.136 An OSH management system that is seen as just a means to get certification will be largely ineffective in its true purpose of continually improving OSH performance via the reduction of work-related accidents and illnesses. The IOSH, in its publication *Systems in Focus* (referred to above), recognises the need for OSH standards and certification processes and recommends the following as best practice in relation to certification:

- Do not allow a business need for external OSH certification to inhibit the development of internal continual improvement processes.
- Do not allow an external certification audit to be viewed merely as a

pass/fail exercise, but rather as an important step along the continual improvement path.

- Develop robust internal OSH management systems before going external.

- Base external audits primarily on evidence gained from internal audit evidence.

- Consider adding external auditor(s) to the internal team (see above) in preference to adding to the number of audits.

- Ensure that all auditors (internal and external) meet the required competence standards (see above).

How to get started on OSH management systems

3.137   In **Chapter 7** we describe how OSH professionals can develop a strategy to enhance their contribution, increase their added value to the organisation and link the OSH risk management system to general risk management and internal controls.

3.138   Set out below is an outline plan on how to get started:

- Choose a system from those discussed above that is best for your organisation. If in doubt, go for the latest version of BS 8800 2004. In the fullness of time, when there is a BS EN ISO standard – more than likely – ISO 18001 – then this will become the preferred option.

- Undertake an initial status review (ISR) – otherwise known as a 'gap analysis'. This will identify those components of any existing OSH system that are working well and will highlight any weaknesses where improvements are necessary. In some cases, the ISR may result in completely new processes being introduced as part of the overall OSH management system. The purpose of the ISR is to ensure that effort is not wasted on reinventing the wheel, especially when the existing wheel is going round OK!

3.139   In **Chapter 7** we describe a process we have developed called a OSH Risk Management Review Tool. This unique tool is used to undertake a top-level risk-profiling review, adopts a holistic approach and looks at the organisation as a complete system. It focuses on the way in which an organisation's risks and the interrelationship between them can have an impact on the management of its health and safety risks and the health and safety function. It reviews four major factors:

- **Factor 1:** health and safety strategic, tactical, operational, professional and personal risks.

- **Factor 2:** organisational context within which the organisation is operating, eg the organisation's activities, where they are based, the ownership status.

- **Factor 3:** policies and systems for general risk management, and other organisational processes.
- **Factor 4:** risks generated by the organisation's business and operational processes, which either have a direct or indirect effect on the management of health and safety risks.

3.140 The methodology seeks to consult as many internal stakeholders – especially the workforce representatives – as is practicable. This then helps to ensure full ownership of and commitment to the resulting findings, recommendations and action plan to enhance the organisation's OSH risk management system.

3.141 The result of the ISR should be an action plan – complete with designated responsibilities and agreed timescales – designed to plug/mind any identified gaps in whatever system/process is currently operational within the organisation. The most successful OSH management systems evolve following the ISR and are not created by it. Rather, they are developed through effective ongoing performance measurement, review and continual improvement.

3.142 It is vital, however, to report the results of the ISR and its associated gap-plugging action plan to the board, senior and line managers and the workforce in order to get the show on the road as soon as is reasonably practicable!

- Make it happen by allocating sufficient resources – time, money and people – to ensure effective and efficient development, implementation and ongoing review of the OSH management system. Techniques used to assist in this regard include:
  - Clear support and personal, visible commitment from the organisation's leaders, especially at board level.
  - Inclusion of OSH targets within high-level business targets.
  - Regular secondment of key managers/professionals to the OSH management system development/implementation team.
  - Customisation of the OSH management system to suit the needs and culture of the organisation.
  - Ensure full workforce consultation, participation, involvement and ownership.
  - Making benchmarking contacts with other like-minded organisations.
  - Having pilot OSH management systems/schemes in selected, friendly areas before going for full system roll-out.
  - Not taking too much time to develop the perfect system prior to implementation, ie adopt the 80/20 rule, as you can always improve, as you get better. Internal audits, management reviews and ongoing action plans will helps to identify required improvements.
  - Recognition and celebration of all successes, irrespective of how big or small their impact is on the overall system.

CONCLUSION

3.143   This chapter has examined the contribution made by OSH management systems to the overall management of risk within organisations. Management systems provide an objective framework that enables organisations to be much more structured and proactive in their risk management as opposed to the piecemeal, reactive systems relied on to achieve the minimum standard of legislative compliance.

3.144   OSH management systems therefore greatly contribute to the ultimate goals of continual improvement and business sustainability and also demonstrate to the organisation's workforce and the outside world – society at large – how the organisation is taking seriously its responsibilities for OSH risk management, corporate governance, business risk management and corporate social responsibility.

CHAPTER 4

Occupational Safety and Health – policy and implementation process

INTRODUCTION

4.1 This chapter follows on logically from **Chapter 2** and **Chapter 3**. Indeed, in **Chapter 3** reference is made to the Health and Safety Executive's publication entitled *Successful Health and Safety Management* – which gave rise to the mnemonic 'POPIMAR' which acts as the framework for this chapter.

4.2 In **Chapter 5** we describe the process for implementing OSH business-related processes. These processes have been drawn from our extensive experience, whilst this chapter provides a legal and overall backdrop to the subject of OSH business integration set out in **Chapter 5**.

4.3 POPIMAR may be used to describe the key stages in the development and implementation of an effective and efficient OSH policy, which should be the cornerstone of any OSH management system. The key stages are:

P	olicy.
O	rganisation.
P	lanning.
I	mplementation.
M	onitoring.
A	udit.
R	eview.

4.4 It is vital that all stages are integrated to create a closed-loop process that feeds back requirements/suggestions for continual improvement to the policy formation and review stages. If the board and senior management are not provided with regular feedback, they will wrongly assume that everything is in order and is going to plan. This is unfortunately the scenario with reactive OSH management, which all too often results in a major accident, before anything constructive takes place (see **Chapter 2**) – too little, too late!

4.5 The POPIMAR approach to OSH policy and OSH management systems clearly demonstrates that the organisation has taken on board corporate governance (CG), corporate social responsibility (CSR) and business risk management (BRM) concepts and practices in their overall management of risk.

POLICY

4.6   Effective OSH policies should set a clear direction for the organisation to follow. This should contribute to all aspects of business performance as part of a demonstrable, visible commitment to continual improvement in OSH management. Responsibilities to the sustainability of people and the environment are met in ways that fulfil the spirit and the letter of the law. Stakeholders' expectations are satisfied. Cost versus risk approaches to preserving and developing human and physical resources, which, in turn, reduce financial and non-financial losses and liabilities should also be incorporated into the policy document.

4.7   OSH policy statements are required by law under the Health and Safety at Work etc Act 1974 (HSWA 1974), s 2(3) and should therefore be prepared and applied throughout the organisation. The most important precursor to preparing and applying a proactive OSH policy is the existence of a positive OSH culture within the organisation. This requires that the avoidance, prevention and reduction of OSH risks at work must be accepted – built in, not bolted on – as part of the organisation's approach and attitude to all its activities. It should be clearly recognised at all levels of the organisation from the boardroom down, through line management to the workforce, and should be prominently reflected in the organisation's vision statement. These aspects are covered in more detail in **Chapter 6**.

4.8   Each organisation should have a unique OSH policy that clearly and unambiguously sets out how OSH is managed within the organisation. It is a unique document that outlines who does what, and when and how they do it.

4.9   The OSH policy statement should be the starting point – and indeed the focal point – for OSH management in the workplace. By law (HSWA 1974); s 2(3), if the organisation employs five or more people, then this policy statement must be in writing.

4.10   The original Health and Safety Executive (HSE) guidance on OSH policies (HSE, *Effective Policies for Health and Safety*, London: HMSO, 1980) recommended that policies should comprise three parts:

- Part 1: General statement of intent.
- Part 2: Organisation – people and their duties.
- Part 3: Arrangements – systems and procedures.

4.11   With the advent of the explicit legal need to undertake workplace risk assessments contained in the Management of Health and Safety at Work (MHSW) Regulations 1992, updated in 1999 (see **Chapter 2**) more recent

HSE guidance implies a four-part policy (HSE, *Stating your Business*, INDG 324, Sudbury: HSE Books, 2000):

- Part 1: General statement of intent.
- Part 2: Organisational responsibilities.
- Part 3: Risk assessments.
- Part 4: Arrangements.

4.12 The simplest Part 1 we have seen simply states:

'It is the policy of this organisation *not* to have accidents.'

4.13 However, in the current climate there is a need for a little more information than that! A more detailed example and explanation are included in **Chapter 5**.

4.14 Writing a health and safety policy statement is much more than just a legal requirement; it demonstrates the organisation's commitment to planning and managing OSH. It is the key to achieving acceptable standards, reducing accidents and cases of work-related ill-health, and it shows all stakeholders that the organisation cares about (their) health and safety. Whatever is written in the policy document has to be put into practice. Hence, the true test of an OSH policy is the actual practices and conditions in the workplace – not how well the document is written!

General statement of intent

4.15 Recent HSE guidance referred to above ('Stating your business') suggests the following format for a general statement:

- To provide adequate control of the health and safety risks arising from our work activities.
- To consult with our employees on matters affecting their health and safety.
- To provide and maintain safe plant and equipment.
- To ensure safe handling, storage, transportation, use and disposal of substances.
- To provide information, instruction and supervision for employees.
- To ensure all employees are competent to do their tasks and to give them adequate training.
- To prevent accidents and cases of work-related ill-health.
- To maintain safe and healthy working conditions.
- To review and revise this policy as necessary at regular intervals.

4.16 The policy should be signed by the most senior person either within the organisation or based on the particular site. It should be dated to facilitate review/revision, which, ideally, should take place at least once every three years. The review date should be shown on the document.

4.17   Once the policy is in place, the legal duty (under HSWA 1974, s 2(3)) is to bring it (and any revision of it) to the attention of all employees. How this is achieved is entirely up to the employer.

4.18   Many ways have been used within industry and commerce to bring the policy to the notice of all employees. These include:

* Full circulation of a paper copy to all employees, possibly asking for a signed receipt.
* Displayed copies on all health and safety notice boards, together with a framed copy in reception.
* Inserts in wage packets/salary notifications.
* Copy logged on the company Intranet/IT networks – this assumes that all employees have 24/7 access to such a system.

4.19   As stated above, the OSH policy document is the cornerstone of any organisation's OSH management system (see **Chapter 3**). Indeed, the HSE guidance, HSG 65 *Successful Health and Safety Management*, previously referred to, states that effective OSH policies set a clear direction for the organisation to follow. They contribute to aspects of business performance as part of a demonstrable commitment to continual improvement. Responsibilities to people and the environment are met in ways that fulfil the spirit and letter of the law. Stakeholders' expectations of the organisation are satisfied via cost-effective approaches to preserving, developing and sustaining physical, financial and human resources, which reduce losses, liabilities, accidents and diseases.

4.20   The following checklist may prove useful in assessing whether your current OSH policy complies with legal and best practice requirements:

* Does a current OSH policy exist for the organisation/location?
* Is the policy up to date (ie not more than three years old)?
* Has the policy been signed (and dated) by a director/senior manager who has (site) responsibility for OSH?
* Does the policy recognise that OSH is an integral and critical part of the business performance?
* Does the policy commit the organisation/location to achieve a high level of OSH performance with legal compliance seen as a minimum standard?
* Does this commitment include the concept of continual improvement?
* Does the policy state that adequate and appropriate resources – time, money, people – will be provided to ensure effective policy implementation?
* Does the policy allow for the setting and publishing of OSH objectives for the organisation/location and/or for individual directors, managers and supervisors?
* Does the policy clearly place the prime responsibility for the manage-

ment of OSH onto line management, from the most senior executive to first line supervision?

- Has the policy been effectively brought to the attention of all employees/agency staff/temporary employees?
- Are copies of the policy on display throughout the organisation/location?
- Is the policy understood, implemented and maintained at all levels within the organisation/location via the use of suitable arrangements?
- Does the policy ensure that employee involvement and consultation takes place in order to gain their commitment to it and its implementation?
- Does the policy require that it be periodically reviewed (at least every three years) in order to ensure that management and compliance audit systems are in place?
- Does the policy require that all employees – including agency staff and temporary employees – at all levels receive appropriate training to ensure that they are competent to carry out their duties?
- Does the policy contain a section dealing with the organisational framework – people and their duties – so as to facilitate effective implementation?
- Does the policy contain a section dealing with the need for risk assessments to be undertaken – both general and specific – their significant findings acted upon, and an assessment record-keeping system maintained?
- Does the policy contain a section dealing with the arrangements – systems and procedures – by which the policy will be implemented on a day-to-day basis?
- Does the policy contain a section dealing with the monitoring/measurement of OSH performance?
- Does the policy contain a section dealing with planning for and reviewing the organisation's policy implementation and overall OSH performance? (Again this may or should be in the 'Arrangements' section.)

4.21   As far as BS 8800: 2004 is concerned (see **Chapter 3**), OSH management is an integral part of improving business performance and high-level, continual performance improvement is cost-effective. This should be reflected in the OSH policy statement. The point is made that legislative compliance should only be seen as meeting the minimum standard requirement.

4.22   The goals in any policy statement should reflect the following:

- Risk minimisation – prevention of injury, ill-health, diseases and accidents.
- People are a key resource; hence the need to promote employee health and wellbeing.

- Adequate and appropriate resources – time, money, people – need to be provided in order to ensure policy implementation.
- There needs to be access to competent, specialist OSH advice.
- Clear OSH objectives should be set, published and communicated.
- The prime responsibility for OSH management lies with line management on an ongoing, day-to-day basis.
- All OSH responsibilities and accountabilities should be clearly understood, implemented and maintained at all levels within the organisation.
- Thorough employee involvement/consultation so as to gain commitment to the OSH policy and its implementation throughout the organisation.
- Appropriate training of all employees to ensure competence – square pegs in square holes!
- Periodic policy review so as to ensure – via management audits – that the policy is up to date and enables the goals of legislative compliance and continual improvement to be achieved.
- Periodic internal and external reporting of OSH performance (see **Chapter 2** for further discussion on OSH reporting in Annual Reports).

ORGANISATION

4.23 This is essentially the 'Part 2' of the vast majority of OSH policies: organisational responsibilities – people and their duties.

4.24 The overall and final responsibility for health and safety rests on the employer. Ideally the person signing the policy has this responsibility on behalf of the employing organisation. HSC's guidance on the responsibilities of directors for OSH management (HSE Books – INDG 343, 2001) clearly states that boards of organisations should appoint one of the board members to be the nominated 'health and safety director'. The guidance also outlines five action points for directors to implement as part of their OSH management systems. These are as follows:

1   The board needs to accept formally and publicly its collective role in providing health and safety leadership in its organisation.
2   Each member of the board needs to accept their individual role in providing health and safety leadership for their organisation.
3   The board needs to ensure that all board decisions reflect its health and safety intentions, as articulated in the health and safety policy statement.
4   The board needs to recognise its role in engaging the active participation of workers in improving health and safety.
5   The board needs to ensure that it is kept informed of, and alert to, rel-

evant health and safety risk-management issues. The HSC recommends that boards appoint one board member to be the 'health and safety director'.

4.25 However, although the overall/final responsibility rests at board level, day-to-day responsibility for ensuring that the policy is put into practice may be delegated to someone lower down the organisation – eg works director/manager – who acts on behalf of the board and who keeps the board fully informed of all OSH-related matters.

4.26 In larger organisations, further delegation may take place on a functional, or area, basis. Each person delegated should be given the responsibility of ensuring that OSH standards are maintained and improved within their functions/areas. Either job titles or names should be incorporated into Part 2 of the policy document. Additionally, specific responsibilities and measurable accountabilities should be included in individual job descriptions, as appropriate.

4.27 All organisations should ensure that all delegated individuals are competent to undertake their OSH responsibilities and are provided with adequate resources – time, money and people – to enable them to do their job properly. It is important that responsibilities and accountabilities are clearly spelled out, together with clear reporting lines. This will make sure that if there are any OSH concerns, they can be reported to the right person so that they can be promptly dealt with. Most Part 2s contain a management structure diagram (or series of diagrams) to show how the delegating down (responsibilities) and reporting up (accountabilities) should happen in practice. An example of an OSH management structure is contained in **Chapter 5**.

4.28 A final paragraph on employees' responsibilities is required to complete Part 2. This could read as follows:

'All employees have to:

- Co-operate with supervisors and managers on health and safety matters.
- Not interfere with anything provided to safeguard their health and safety.
- Take reasonable care for the health and safety for themselves and others.
- Report all health and safety concerns – accidents, hazards, defects, shortcomings, near-misses – to an appropriate person, as detailed in this policy statement.'

4.29 From an organisational viewpoint, knowledge of responsibilities and accountabilities, coupled with positive and proactive working relationships, is an excellent way of promoting a positive OSH culture. In order to get the necessary level of Commitment (the fifth C) throughout the organisation, it is imperative that use is made of the (other) four Cs:

- Control.
- Co-operation.
- Communication.
- Competence.

4.30   Control is achieved by securing the commitment of all employees to clear OSH objectives. Managers must take full responsibility for controlling all those factors, which may lead to injury, disease, damage or death. They should provide clear direction to their staff and also ensure that a safe and healthy working environment is provided at all times for all employees, irrespective of when and where they are working – eg third-party premises, abroad, driving, etc.

4.31   In order to achieve these high and continually improving standards of OSH management, control must be established and maintained. OSH responsibilities and accountabilities must be allocated to line managers whilst specialists – competent persons – act as their advisers. A senior figure – ideally a main board director – should be publicly nominated as the director responsible for OSH management policy implementation, co-ordination and monitoring (see the HSC list of directors' responsibilities, above).

4.32   The key control measures that should be in place comprise:

- Policy and organisation – people and their duties.
- OSH objectives, targets and their review.
- Planning, monitoring and auditing OSH activities so as to ensure legislative compliance and risk minimisation on a global basis throughout the organisation.
- Effective and efficient implementation of individual and collective action plans.
- Written job descriptions with clear accountabilities and key performance indicators (KPIs).
- Performance standards: systems, procedures, rules.
- Formal performance review and appraisal systems.

4.33   Co-operation involves getting participation, commitment and involvement with OSH activities at all levels within the organisation. This is an essential prerequisite in any effective risk-control system. This is best achieved via a proactive OSH culture, which has 'ownership' as its cornerstone. Ownership of policies, procedures and practices makes OSH everyone's business!

4.34   Key factors include:

- Active encouragement and support of appointed workplace safety representatives by clearly recognising the valuable contribution their role makes to the continual improvement process.
- Group involvement and participation in setting performance standards

via task groups, OSH circles, focus groups and the like. Experience has shown that much tighter and rigorous standards are developed by such means, when compared to those merely imposed by management following little or no participation/involvement of employees.

- Auditing teams – using a mix of internal and external auditors (see **Chapter 3**).
- Ad hoc problem-solving teams drawn from the working population exposed to the problem.
- Provision of specialist support (competent persons) to assist in the problem resolution.
- Use of hazard/near-miss reporting logs to enable employees to fulfil their legal requirement to report any 'shortcoming in the employer's protection arrangements for health and safety' (MHSWR 1999, reg 14).

4.35 Communication is essential in any OSH management system; information should flow into, within, and out of the organisation.

4.36 *Into* the organisation flows OSH information and intelligence, which is necessary to monitor legal, technical and managerial compliance with new developments and initiatives, so as to ensure that all aspects of the OSH control systems are valid and up to date.

4.37 *Within* the organisation, communication may be via:

- Visible behaviour – setting a good example.
- Written communication.
- Face-to-face discussion:
  - Safe visiting.
  - Toolbox talks.
  - Team briefing.
  - Training, both formal (classroom) and informal (on the job).

4.38 *Out of* the organisation includes:

- Product safety literature.
- Material safety data sheets (also *Into* the organisation).
- Liaison with emergency services, local authorities (MHSWR 1999, reg 9).
- Informing the local community/general public of the potential impact on them from work activities.

4.39 Competence of all employees – including management and specialists – must be ensured and maintained so as to maximise the overall contribution to the continual improvement of the OSH management system. Placing competent square pegs in squares holes is also a legislative requirement (MHSWR 1999, reg 13) as is the need placed on the employer to appoint competent persons to assist 'in undertaking the measures he needs to take to comply with the requirements (dos) and the prohibitions (don'ts) imposed upon him by or under the relevant statutory provisions'.

4.40   These required levels of competence might be achieved via suitable:

- Selection, recruitment and placement procedures:
  - Capabilities (physical and mental) for the job.
  - Training/skills/experience.
  - Testing/examinations (ability/medical).
  - Induction.
- Identification of training needs:
  - Training needs analysis.
  - Changing situations.
  - Refresher training.
  - Specialist training needs.
  - Contractors/visitors.
- General provision of necessary information, instruction, training and communication for employees.
- Sufficient cover for absences, especially in the case of persons having OSH-critical responsibilities.
- General health and wellbeing promotion and surveillance.
- Meaningful hands-on experience (including role play) under proper supervision.
- Knowledge of and working in compliance with agreed performance standards.
- Provision of – or access to – competent OSH advice, guidance and support.

4.41   Once the four Cs have been brought together and incorporated into the OSH management system, the fifth C – Commitment – will then be able to flourish and hence the goal of continual improvement attained.

4.42   As stated above, clearly defined and agreed responsibilities are a necessary prerequisite in the 'Organisation' section of any OSH policy document. These responsibilities may be divided into three subsets:

1   Managerial.
2   Organisational.
3   Individual.

4.43   Managerial responsibilities include the fact that the ultimate responsibility for OSH management rests with the board of directors (see above) on both an individual and collective basis. There is also a need – throughout the organisation – to ensure that OSH responsibility, accountability and authority have been clearly delegated and accepted. In turn this will result in:

- Ensuring that the OSH management system is developed, implemented, and is performing well throughout the organisation.

- Appointing sufficient competent persons to assist (see above).
- Achieving all set OSH objectives and targets within agreed timescales.
- Regular reviews and evaluation of the efficiency and effectiveness of the OSH management system.
- Regular, periodic OSH performance reporting/communication upwards to board level and outwards to stakeholders via the Annual Report (see **Chapter 2**).

4.44 Organisational responsibilities should ensure that people at all levels within the organisation are competent, trained and fully aware of the relevance and importance of their individual and collective OSH responsibilities/activities. This may involve specific, functional responsibilities being set and agreed for functions such as human resources (HR)/personnel, training and development, purchasing, OSH practitioners, hygienists, ergonomist, occupational health practitioners, etc.

4.45 Specifically the OSH/HR agreed training programme – which is a fundamental organisational responsibility – should be developed and implemented so as to ensure that all employees are:

- Held responsible for the OSH of those they manage and work with, and of themselves.
- Made aware of OSH responsibilities towards others who may be affected by what they do and/or don't do – ie acts and omissions.
- Made aware of the influence that these acts and omissions may well have on the overall performance and effectiveness of the OSH management system.
- Made aware of the 'mechanism of harm' from hazards in the workplace, especially in connection with disease/ill-health causative patterns.

4.46 Individual responsibilities should therefore:

- Be clearly allocated and communicated to all concerned.
- Be clearly defined via job descriptions.
- Have authority and resource provision – including time – clearly assigned.
- Have appropriately accountability fixing arrangements – eg regular feedback reporting.
- Have clear, unambiguous lines of command/reporting arrangements; the use of a management structure diagram – an organisational organogram – will greatly assist in this regard and should be incorporated into the policy document.
- Have OSH performance incorporated into the appraisal system via the setting of at least two or three OSH key performance indicators (KPIs) for every employee who is in control of people, premises, processes, purchases, products or pounds (the six Ps).

These aspects are also discussed in **Chapter 5**.

4.47 Finally, within the 'Organisation' section of the OSH policy document, there is a need for clear organisational arrangements to be in place – as part of the OSH management system – in order to deliver the following:

- Access to OSH knowledge, skills and experience via the recruitment and appointment of competent OSH practitioners.
- The ability amongst all employees to identify, eliminate and/or control OSH hazards in the workplace.
- The promotion of health and wellbeing at work via the minimisation of occupational and non-occupational health hazards in the workplace.
- A clear definition of responsibilities and accountabilities throughout the organisation using an organogram (see above).
- People with skills and authority to live up to their roles and responsibilities.
- Training/communications programmes that bridge/mind the gaps in existing skills and knowledge.
- Effective employee involvement, consultation and representation.
- Employees committed to good, continually improving OSH performance.
- Effective communication – into, within and out of the organisation – of OSH information.
- Effective liaison and communication with external agencies, such as enforcers, insurers and emergency services.
- Sufficient, suitable advisory and support services to enable line management to fulfil their OSH responsibilities and objectives.

PLANNING

4.48 'Failure to plan is planning to fail!'

There should be a planned and systematic approach to implementing the OSH policy in order to act as the foundation of an effective and efficient OSH management system. Without a formalised plan – whether it be an annual, three-year or five-year plan – there can be little or no proactive progress towards the goal of continual improvement.

4.49 The 'OSH director' should own the plan, and progress on the plan should be regularly reviewed and the results communicated to all concerned.

4.50 There should be agreed actions, responsibilities, timescales and accountabilities built into the framework of the plan via the use of key performance indicators (KPIs) and performance appraisal interviews (one-to-ones) between the OSH director and the accountable managers/supervisors.

4.51 Effective planning should therefore result in an OSH management system that controls risks, reacts to changing demands – both internal and external to the organisation – and sustains positive OSH culture.

4.52 Planning involves designing, developing and installing risk control systems (RCSs) and workplace precautions (WPs) commensurate with the organisation's identified hazards and assessed risks. It also involves operating, maintaining and improving the OSH management system to suit ever-changing needs, hazards and risks.

4.53 The planning process should address the following key areas:

- Setting objectives.
- Risk-control systems.
- Legal and other requirements.
- OSH management arrangements.

Setting objectives

4.54 In order to avoid duplication of effort, it is vital that the success and/or failure of all planned activities within the OSH management system are transparent, understood and thoroughly communicated.

4.55 The OSH objectives to be contained in the plan should be drawn up following the initial baseline audit – the initial status review (ISR) – that should have answered the question 'Where are we now?' in terms of the existing (if any!) OSH management system. Once the system is up and running, new objectives may be set following audit and periodic review (see below).

4.56 As stated above, clear unambiguous performance indicators and measurement criteria defining what is to be done, by whom, by when, and with what outcome, should be set.

4.57 The key elements in planning and setting objectives may be summarised as follows:

- All objectives should be specific to the organisation and should be appropriate relative to its size, spread, nature of activities/processes, the hazards and the conditions in which it operates.
- All objectives should be clearly defined, quantified/qualified and prioritised.
- All objectives should have as their focus the continual improvement of the OSH management system.
- Suitable measurement criteria should be selected in order to confirm that the objectives are being met in practice.
- A plan should be drawn up that is designed to achieve each objective.
- Adequate time, money and people resources should be made available; this should also include an allowance for any technical/competent person support needed.

- The implementation of the plans and their efficiency and effectiveness in achieving the desired objectives are measured, reviewed and communicated.

Risk control systems (RCSs)

4.58   It is essential in designing RCSs to ensure that all risks encountered are reduced to tolerable or acceptable levels so far as is reasonably practicable. The use of a Risk Register and the identification of a risk profile for the organisation are necessary prerequisites in this regard. These matters are discussed in **Chapter 7**.

4.59   The planning process should clearly define the arrangements for:

- Ongoing, proactive hazard identification and assessment of OSH risk arising out of the work activities and the working environment.
- Development and implementation of effective and efficient RCSs and workplace precautions (WPs) that eliminate hazards or minimise risks.
- Recording the significant details and findings of each risk assessment, irrespective of the degree/level of risk involved.

4.60   A risk-based control plan should be developed that encompasses different risk levels – both significant and insignificant/trivial – and also considers risk tolerability and action timescales. The following is taken from BS 8800: 2004, Annex E.

Risk	Tolerability/action timescale
Very low	Risk acceptable
(Insignificant/trivial)	No further action required Maintain existing controls
Low	Risk regarded as tolerable No additional controls, unless cost is very low Maintain existing controls Further risk reduction – low priority

Medium	May or may not be tolerable
	Risk should be reduced to tolerable/acceptable
	   levels, but cost of control measures should be
	   taken into account
	Implement risk-reduction measures within
	   defined time period
	Maintain existing and new controls

Risk	Tolerability/action timescale
High	Only tolerable in rare cases Maintain existing and new controls Substantial efforts should be made to reduce risk Urgent implementation of risk-reduction measures Consider suspending/restricting activity Consider resource implications
Very high	Risks are unacceptable Substantial improvements in risk controls essential Risk must be rapidly reduced to acceptable/tolerable level Work activity must be prohibited/halted until risk controls in place If risk cannot be reduced to tolerable/acceptable levels, the activity should be discontinued

Legal and other requirements

4.61 The organisation should establish and maintain arrangements – usually via the employment of competent OSH practitioners – to ensure that all current and emerging OSH legal and other standards/requirements are understood and are being complied with, bearing in mind that best practice assumes that legislative compliance only equates to the minimum standard.

4.62 Benchmarking within the organisation's employment sector can ensure best practice and performance measures are adopted and communicated. The HSE's guide – INDG 301 (1999) – is entitled *Health and Safety Benchmarking – Improving Together* and sees the process as a five-step cycle aimed at ensuring continual improvement:

- Step 1: Decide what to benchmark.
- Step 2: Analyse where you are now (initial status review/baseline audit).
- Step 3: Select your partner.
- Step 4: Work with your partner.
- Step 5: Act on the lessons learned.

4.63 OSH benchmarking produces the following advantages:

- Improves the organisation's standing/reputation.
- Avoids reinventing the wheel by learning from the experience of others' good ideas by comparing and contrasting how things are done.

- Develops relationships with key stakeholders, including customers, suppliers and contractors.
- Enables organisations to establish where they are now and, more especially, where they are in the league table – promotion/relegation?
- Saves money and maintains the organisation's competitive edge; savings may be realised via lower insurance premiums, increased productivity and reduced staff turnover.
- Improves the overall management of OSH, thereby reducing risks to people's health and safety.

OSH management arrangements

4.64   The OSH management arrangements (the Part 4 of the OSH policy and a requirement of the MHSWR 1999, reg 5) should reflect all the needs of the organisation wherever and whenever it is operating and should also reflect its risk profile. The arrangements should therefore cover all elements of POPIMAR; this is especially important to ensure smooth implementation.

4.65   Therefore, the following key areas should form part of the OSH management arrangements:

- Plans, objectives, people, financial and time resources.
- Operational plans to develop RCSs and WPs in order to reduce/minimise workplace OSH risks.
- Contingency plans for disasters/emergencies that incorporate effective mitigation considerations.
- Plans for organisational activities (see above).
- Plans covering aspects of change management.
- Plans for interaction with interested stakeholders such as contractors.
- Plans to cover performance monitoring, audit and review (see below).
- Methods of ensuring that corrective actions are implemented in a prioritised and timely manner – eg action plans (see above).

IMPLEMENTATION

4.66   This element is the most important phase of POPIMAR. The POP stages have set the scene; the Implementation phase is when things start to happen, actions are taken and completed, and the organisation begins to adapt, adopt and improve its OSH management system which starts living up to its aim of minimising risk throughout the organisation. Risk profiling and risk assessment should be used to decide on priorities and to set objectives – via the plan (see above) – for eliminating hazards and reducing risks within the workplace irrespective of what and where they may be.

4.67   Performance standards should be established and used for measuring and demonstrating achievement. Specific actions designed to promote a positive OSH culture should be built into the implementation process. The following is a suggested list of subject areas, which may need to be included/audited as part of the implementation process:

- Business risk management profiling review.
- Recruitment, selection and training.
- Consultation and communication.
- Risk assessments.
- Control of hazardous substances.
- Noise.
- Manual handling.
- Display screen equipment.
- Personal protective equipment.
- Safe systems of work/permit to work systems.
- Management of third parties on site: contractors/visitors.
- Electrical safety.
- Workplace safety.
- Work equipment safety.
- Fire/emergency arrangements.
- First aid arrangements.
- Occupational health and welfare considerations.
- Business and operational process change.
- Work related driving.

4.68 Risk-control systems, workplace precautions and management arrangements are much easier to implement if they are well designed and can be seen to be building onto existing, accepted business practices. Documentation is a key element in enabling an organisation to communicate about and effectively implement an OSH management system. It also greatly assists in assembling and retaining OSH knowledge and skills within the organisation.

4.69 OSH documentation should be:

- Kept to the minimum required for effectiveness and efficiency – poor systems suffer because of information overload and over-bureaucracy!
- Tailored to suit the organisation's needs.
- Detailed proportionate to the level of complexity, hazards and risks.

4.70 Amongst the most important written communications are:

- OSH policy statements.
- Organogram showing OSH roles, responsibilities and accountabilities.
- Risk assessments:
  - General and specific.
  - Health as well as safety.
- Documented performance requirements, current year objectives (CYOs) and key performance indicators (KPIs) for individuals and groups.

- Organisational and risk control procedures and information.
- Training records:
  - General and specific.
  - Formal and informal.
  - Individual.
  - Syllabi, course duration, attendees.
  - Whether training is accredited/examinable or not.
- Findings and recommendations from baseline audit/initial status reviews.
- Preventive measures following accident investigations.
- Results of periodic (annual) status reviews.

4.71 It is important that written OSH records are available for use by the organisation and its people on a 24/7 basis. This may require both paper and electronic versions, especially if there is any doubt as to whether access to the system can satisfactorily be gained by a process worker on a Sunday night shift!

4.72 The organisation should therefore maintain and keep up to date sufficient records to:

- Demonstrate compliance with legal and other requirements.
- Ensure retention of relevant and appropriate OSH knowledge.
- Mitigate any liability claims – this is especially important in the light of the Woolf reforms on civil liability which set tight timescales for the production (to the plaintiff's solicitor) of a whole raft of OSH-related documentation.
- Provide sufficient data on which to base future OSH plans and initiatives.

MONITORING

4.73 OSH performance should be monitored/measured against agreed and accepted (by the organisation) standards in order to reveal when and where improvements are needed.

'What gets measured gets done!'

4.74 There are two types of monitoring required: active and reactive.

1   Active monitoring – before the event – reveals how effectively (if at all) the OSH management system is functioning. There is a need to examine both the hardware – premises, plant/equipment, and substances – as well as the software – people, processes, procedures and systems. This usually involves a mix of audits, surveys, inspections, tours, safety sampling, safe visiting and OSH task or focus groups.

2   Reactive monitoring – after the event – is in effect an analysis of fail-

ures in the system, ie when and where the RCSs and the WPs have not operated as designed, or do not even exist! The use of accident investigation techniques and numerical/causal statistical analyses should help to highlight why the system has gone wrong and should guide management towards system improvements. All accidents – injury, disease, damage, near miss – should be used as learning events and warning signs that all is not well.

4.75 A positive OSH culture does not allow fault-finding and blame apportionment to cloud the issues and thereby prevent the implementation of actions designed to prevent a recurrence of the accident under investigation. The objectives of reactive monitoring are therefore to determine the immediate causes of substandard performance, to identify the underlying causes (of substandard performance) and hence the implications for the future design, planning and operation of the OSH management system and its continual improvement.

4.76 Hence, the primary purpose of performance monitoring is to identify achievements and measures that proactively prevent failure. It should also provide information on the progress and current status of the arrangements – strategies, processes and activities – used by the organisation to control OSH risks.

4.77 Such monitoring information sustains the operation and development of the OSH management system and hence the control of risk via the:

- Provision of information on how the system is actually operating in practice.
- Identification of those areas where specific remedial action is required.
- Provision of a solid base on which to build towards continual improvement.
- Provision of feedback and motivation in order to promote the desire to do things better, safer and healthier.

4.78 OSH performance monitoring data should ideally provide answers to the following questions:

- Where are we now relative to the organisation's objectives?
- Where are we now in terms of identifying hazards and controlling risks?
- How do we compare with others, ie benchmarking?
- Why are we where we are?
- Are we getting better or worse over time?
- Is our OSH management system effective, ie doing the right things?
- Is our OSH management system efficient, ie cost-beneficial?
- Is our OSH management system reliable, ie doing the right things consistently?

- Is our OSH management system proportionate to the organisation's hazards and risks?
- Is the OSH management system effectively in place and operational throughout the organisation?
- Is the organisation's culture supportive of OSH management, particularly in the face of competing demands for scarce resources?

4.79 The above questions should be asked – and positively answered – at all management levels throughout the organisation. All managers should be given the responsibility, authority and accountability for OSH performance monitoring in those areas/functions for which they are responsible and 'own'.

4.80 Key performance indicators (KPIs) should be set at/for all levels of management from the top down, with accountabilities being fixed via regular cascade-type reporting from the bottom up. Such KPIs should be a mix of qualitative and quantitative measures and should also reflect both proactive and reactive measures. As stated above, proactive indicators may involve monitoring compliance to agreed standards via surveillance, inspections and adherence to systems and procedures. Reactive indicators may involve monitoring accidents, near misses, ill-health and other historical failure data as evidence of deficient OSH performance.

4.81 Although the primary focus for performance monitoring is to meet the internal needs for organisational OSH data, there is an increasing need to demonstrate to external stakeholders (see **Chapter 2**) that arrangements to control OSH risks are in place, are operating correctly and are effective. As stated above, organisations should clearly and publicly communicate details of their OSH performance – good and bad – to all their stakeholders.

Substandard performance

4.82 Arrangements should be in place so as to ensure that there is a consistent and objective response to – and a thorough investigation of – all cases of substandard performance. Such investigations should be proactive in nature and should never become fault-finding, blame apportioning exercises! The results of these investigations should be analysed and reviewed in order to learn from the warning signs and to identify any common features and/or trends, which should lead to the development of commensurate control measures/corrective actions. The use of competent investigations will greatly assist in this regard.

4.83 The level, nature and depth of the investigation should reflect the significance – potential rather than actual – of the events and their results and should identify:

- Reasons for substandard performance.
- Immediate causes for OSH management system failure(s).
- All underlying root causes and any learning events.

- Any recommendations for changes in existing systems/procedures and/or the introduction of new systems/procedures.

4.84   Investigations are also needed to:

- Satisfy notification, recording and reporting requirements (eg RIDDOR).
- Collect information needed in the event of legal action.
- Collect information for insurers in connection with potential claims.
- Maintain organisational records.
- Identify and rectify immediate and underlying causes.
- Make recommendations for remedial actions in order to further improve the OSH management system both in the short and medium term.

AUDIT

4.85   The organisation should audit all elements of the OSH management system in order to learn from all relevant experiences and thereafter should apply the lessons learnt in order to move towards the continual improvement of the system.

4.86   There should be systematic internal and external audits in order to examine OSH performance data from both proactive and reactive sources (see above). Such auditing not only ensures legislative compliance – the accepted minimum standard – but also engenders a strong individual and organisational commitment to the constant development of policies, systems and techniques of OSH risk control. Performance – both individual and collective – should be assessed via audit against agreed internal KPIs and external benchmarking, and the findings and recommendations for improvement implemented on a timely basis and communicated to all concerned (see **Chapter 3** for further discussion on typical audits).

4.87   In essence – rather like a comparable financial audit – the OSH audit is a deep and constructively critical appraisal of all the elements (POPIMAR) of the OSH management system. It should be seen as an additional exercise to the routine monitoring described above and also to the initial (ISR) and periodic (PSR) status reviews.

4.88   The auditors need to be competent and independent of the area/activity being audited (see **Chapter 3**) and they should be tailored to suit the needs/hazards/risks of the organisation.

4.89   The scope of the audit should inter alia be able to answer the following questions:

- Is the overall OSH management system capable of achieving the required levels of performance?
- Is the organisation living up to all its OSH obligations?

- What are the current strengths, weaknesses, opportunities and threats within the OSH management system? (SWOT analysis)
- Is the organisation actually doing and achieving what it claims? (Mind the gap(s)!)

4.90   Audits may be broadbrush/comprehensive, covering the full range of the organisation's locations and activities, or they may address certain key, selected subject areas. In all cases it is imperative that the audit results and corrective actions are widely communicated, and clear responsibilities and accountabilities fixed for audit report implementation and follow-up, ie who is going to do it, by when, and who is going to check to see that corrective actions have been satisfactorily completed on time.

4.91   Senior management should use any current, relevant audit report during the 'Review' procedure (see below).

4.92   There are many proprietary audit systems available throughout the worldwide OSH community. Some are qualitative; most are quantitative and involve a 300–600 question set, in addition to discussions, interviews and walk-round site observations. The best audit system is the one that is accept-able throughout the organisation and the one that fits the organisation's risk profile and OSH culture. The best audit system is the one that gets used!

REVIEW

4.93   There are three different types of review procedures:

1   Initial status review (ISR).
2   Periodic status review (PSR).
3   Management review (MR).

4.94   The ISR has been described above and, in essence, answers the question 'Where are we now?' in terms of the existing OSH risk-management system.

4.95   The PSR is usually undertaken at least quarterly and is used to review the status of the organisation's current OSH plan. Judgments should be made concerning the adequacy of OSH performance against plan, in order to enable decisions to be made about the nature and timings of required remedial actions necessary to improve the overall system and plug any gaps.

4.96   Hence, the PSR should consider:

- The overall performance against plan of the OSH management system.
- The performance of each of the individual elements – POPIMAR.
- The findings of all audits since the last PSR.
- Internal and external (to the organisation) factors, such as changes in organisational structure/personnel; production/shift patterns; pending legislation, codes and guidance; introduction of new technology/

processes; acquisitions, mergers and disposals; managing new contracts, clients, suppliers, distributors, etc.
• Anticipated future changes.

4.97   The PSR should be based on information gathered via proactive and reactive monitoring, audit findings and accident investigations. Ideally, this information gathering should be a continual process which should include line management/supervision responses to system failures, such as the non-implementation of agreed risk control systems (RCSs) or workplace precautions (WPs); responses to remedying substandard performances by either groups and/or individuals – this ideally should involve mentoring and counselling, as opposed to fault-finding and blame apportionment. The PSR should also involve all levels within the organisation in the collective discussion, participation and assessment of existing and future plans to improve the OSH management system.

4.98   The MR should be undertaken by the organisation's board and/or top management on a regular – at least annual – basis in order to ensure the continuing suitability, adequacy, efficiency and effectiveness of the OSH management system and its associated policies and objectives. The MR should be based on audit findings – internal and external, performance monitoring reports, and other data that enable valid judgments to be made concerning the appropriateness of the existing OSH policy. This in turn may well lead to the establishment of new or re-validated OSH objectives to be incorporated into future (annual) OSH plans that are focused on continual improvement. This may include changes in existing OSH arrangements.

4.99   The findings of the MR should be documented and communicated so as to ensure appropriate actions are taken throughout the organisation. They should also be incorporated within OSH performance reports for communication to all concerned.

4.100   Following the requirements of Turnbull and other influences on corporate governance and internal control – see **Chapter 2** – it is vital for all organisations to critically examine their policy and management system to check that all OSH hazards and risks have been identified, assessed, controlled and communicated, not only to the workforce but to all interested stakeholders, both internal and external to the organisation.

CONCLUSION

4.101   The best OSH policy is one that is implemented in an integrated manner, that fits the requirements of the organisation, works in practice and one to which the organisation is visibly and demonstrably committed at all times, in all operations, in all locations, and at all levels. This is a statement that could be described as 'easy to say, but difficult to deliver'. The next chapter starts the process of providing guidance on how to make it happen.

CHAPTER 5

OSH business-related processes

INTRODUCTION

5.1 In the preceding chapter we set out general guidance for the development and implementation of an OSH policy and management system. In this chapter we provide specific advice based on our experience of developing such systems in a variety of organisations.

5.2 As mentioned before, if OSH professionals are to get on the 'business and risk agenda', then they need to speak a business language and adopt a risk managing and not legal compliance risk-averse approach.

5.3 Earlier chapters have discussed key building blocks in the process to make OSH part of business risk management. In this chapter we have concentrated on the OSH business processes that we have found can make a big difference to the successful management of OSH within an organisation's 'business and commercial environment' and business and operational processes. In **Chapter 7** we will discuss how OSH professionals can improve their chances of increasing their influence on their organisation or clients, using these processes and others within a business context, to build a strategy for enhanced added value.

5.4 OSH professionals need to view their organisation, or client organisations, as a complete system so that implemented business processes complement one another and are designed to ensure an integrated, consistent and non-duplicating approach. Our experience shows that this approach is appreciated and welcomed by organisations who are generally looking for flexibility, added value and not uncoordinated 'red tape'. They respond much better to the use of business and commercial focused interventions, and can see the added value of good/best practice if it is explained in business terms.

5.5 Our general approach therefore is not to rely on extensive references to statutes or regulations to explain or justify why organisations should consider the implementation of the following business-related processes as part of their contribution to the management of risk. We believe that the OSH business-related processes set out below are required by all organisations. This is for several reasons:

- In general terms, legal requirements are normally set at what we call the 'lowest common denominator level'. That is to say that because government has to justify the introduction of a regulation – particularly to the private sector – it now undertakes a regulatory impact analysis to show the cost – direct/indirect – of the introduction. The tendency is for government to restrict the impact/cost of the introduction, so that the new regulation describes the changes required, but in a way that will only require those organisations with very poor arrangements to take action and incur a cost. Consequently, organisations with even a reasonable approach to OSH management will in general be adopting arrangements already in excess of legal requirements, prior to their introduction.

- We have found that quoting statutes or regulations to already hard-pressed managers/employees, or seeking to justify a course of action by quoting statutes or regulations, is guaranteed to achieve one end result – a completed 'turn-off' for managers/employees. Whilst earlier chapters have referred to statutes and regulations (it is necessary to do so to put the OSH agenda into context), the reality is that if an organisation (in general terms) adopts good/best practice then their OSH management system will be far in excess of legal requirements, and much more effective and efficient.

- Even where an organisation has to make enhancements to its OSH arrangements to satisfy new legal requirements, OSH professionals should adopt a mind-set that views the design, implementation and maintenance of new policies/strategies, etc, as an opportunity to improve the management of the business rather than an opportunity to increase 'red tape', and increase their control.

- OSH professionals should use a balanced cost-benefit approach to show the cost of an intervention and the cost of non-intervention, rather than using new regulations to justify action.

- New regulations should result in business-focused and commercially relevant arrangements that satisfy regulations, but are essentially consistent with, and integrated into, the organisation's day-to-day activities, so any changes required can be managed by managers/employees with the minimum of disruption.

5.6   In our experience, the OSH business-related processes set out below are required by *all* organisations, to a greater or lesser extent. The extent to which they are required will depend on a number of factors – in general terms the hazards/risks of the organisation, and specifically what it does, where it does it, size, ownership structure, management style, background, commercial situation, etc. Organisational factors are discussed in **Chapters 6 and 7**, so we will not go into any detail here.

5.7 For the purposes of this chapter we have assumed that a typical organisation is one that has the basics of an OSH system in place, but a system that is not part of normal organisational and operational processes, and relies on a reactive approach to problems that arise. What 'drives' the organisation are commercial results, to the exclusion of virtually everything else, and OSH and business risk management systems are 'nice to haves', when there is the time and resources available. But not this week/month/year!

ORGANISATION STRUCTURE FOR OSH

5.8 An organisation needs to have an effective policy to describe its overall intentions in relation to the management of OSH, and how it will fit within the organisation, and business risk-management systems. **Chapter 4** has described the key building blocks using the P O P I M A R letters, and we do not intend to repeat those details. However, before thought is given to writing a policy statement – to be signed by the most senior person in the organisation – we recommend that the organisation structure for OSH management be 'described' using an organisational chart or similar. This will result in the most senior person being identified, thereby defining who should sign the policy statement. There is also a need to define the OSH management structure separate to the main organisational structure, as the latter is based on business reporting lines and relative levels of position for normal management processes. Also, where the main structure is confused by different activities being managed at, or from, a particular site, there is a need to define who has overall responsibility for OSH covering the whole 'site' or business unit. Additionally, the line-management function for managing OSH must be defined separate from the supporting/adviser functions. As the management of OSH crosses all these aspects, there is a need to create a totally separate structure that will show the OSH reporting lines, and assist in the allocation of responsibilities, job description inserts and performance measures, etc. OSH risk management is not just about managing the tangible 'hazards/risks', but also the intangible, eg factors that relate to organisational responsibilities, or people management policies. Creating a specific separate management structure for managing OSH recognises that there is a risk of confusion being created as to who has responsibility for OSH, leading to uncertainty and often inaction.

5.9   A typical OSH structure showing OSH responsibilities and the normal organisational position of the main people involved is set out below:

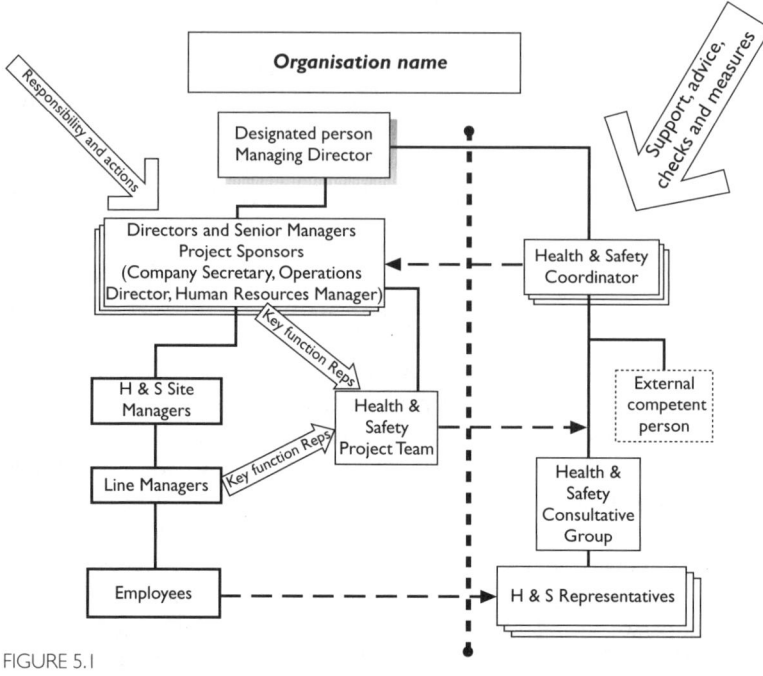

FIGURE 5.1

5.10   In this example, the designated person with overall and ultimate responsibility for OSH is the Managing Director and the OSH Co-ordinator is the Human Resources Manager. The Operations Director, who is part of the Directors and Senior Managers group, is specifically excluded from membership of the OSH Project Team, so that the Project Team can discuss and make recommendations without initially being constrained by resource restrictions. The Human Resources Manager is not normally part of the Directors and Senior Managers Group, but is now invited in to discuss OSH and human resources issues. The OSH Project Team is chaired by the OSH Co-ordinator, and is also attended by the external competent person. The Project Team comprises middle managers from each site (five sites in total), and other managers, who have been carefully selected so that each 'hazard group' is represented – 'offices', external sales, distribution and transport, warehousing, purchasing, showrooms, retail.

5.11   Several of the sites have more functions at the site than the main function, eg group purchasing is based at one of the sites, group customer services at another, and external sales at another, who each report under the main structure to a person at another site.

5.12   To ensure effective control and co-ordination, each site has appointed a manager with specific OSH responsibilities for the total site. The person is responsible for OSH co-ordination for all functions operating out of, or at, the site. This ensures that OSH is not left alone by each function, thinking it is the responsibility of the 'others' on the site, and there is a single focus for OSH on the site.

5.13   OSH representatives have been appointed at each site and covering all 'hazard groups'. The Human Resources Manager chairs the OSH Consultative Group to ensure a consistent approach between the Group and the Project Team that they also chair. The Human Resources Manager consequently has a direct 'OSH reporting line' to the Managing Director. Normally the Human Resources Manager reports to the Company Secretary.

5.14   This example shows how one organisation can create an OSH structure separate to its main structure, reporting lines and relative levels of position that provide a valuable focus for OSH, but at the same time allocates responsibility where it can be most effective. However, one structure does not make a system, as managers and employees need direction, advice, guidance and information in order to undertake their responsibilities. As we have seen from earlier chapters both managers and employees have legal responsibilities and these also need to be built in to OSH and business processes. These processes are discussed below.

OSH BUSINESS-RELATED PROCESSES

5.15   The following processes will be described and discussed:

1   Management system.
2   Communications.
3   Recruitment, selection, learning and training.
4   Risk assessment.
5   Rules and procedures.
6   Health and hygiene.
7   Property conservation, business continuity and emergencies.
8   Hazard control.
9   Personal protective equipment.
10   Planned preventive maintenance.
11   Business process and operational changes.
12   Work equipment and complex machinery.
13   People and vehicle risks.
14   Accident management.
15   Monitoring and review.

Management system

5.16   In general terms, a management system for OSH must be integrated with the normal business processes. The management system should cover:

- Statement of intent (policy).
- Management structure.
- OSH objectives.
- Responsibilities.
- Levels of authority.

Statement of intent

5.17   In our experience, the creation of a 'single page' 'Policy' that is signed by the designated person within the organisation (see above) is extremely useful in a number of ways:

- It is a clear statement of intent, rather than a long series of policies and procedures that are often referred to as the 'Policy', but which tend to be inaccessible to most people, and difficult to understand.
- It is easy to understand and is clearly signed by the designated person.
- Can easily be reviewed on a regular basis.
- Can easily be displayed at all locations.
- Can be given to new employees at induction.
- Can be referred to during management training courses.
- Can easily be provided to external parties, eg contractors, customers.

Management structure

5.18   As described in the previous section, the organisation should create an OSH management organisational structure to reflect the main structure, but also to link to other business risk management systems.

OSH objectives

5.19   These can be quantitative or qualitative. Often organisations believe that quantitative objectives – for example, reducing the incidence of certain negative incidents, such as accidents within a set period – is the most effect-ive measure, as it is related to the focus on quantitative measures that exist within the rest of the organisation, eg sales objectives or reductions in costs. However, as the causes of accidents and their consequences do not have a simple cause and effect relationship (neither have most other business meas-ures – but this obvious fact is normally ignored), it is inappropriate to measure the progress of an OSH management system using 'consequence' measures alone. Consequences are only an indicator, not a measure. What are needed are objectives that relate to progress towards the implementation of an

OSH management system, and the work required to make that progress a reality – what we call 'input' measurements, not 'output' measurements.

5.20 Additionally, 'input' measurements can be made 'SMARTT' (specific, measurable, achievable, realistic, timed and trackable) and can be cascaded down the organisation, and allocated where the 'input' is most needed. Measurements based on 'input' can also be linked to performance management and reward systems (see below).

Responsibilities

5.21 There is a need to specifically allocate responsibilities for the management of OSH, so that managers and employees realise that it is part of their normal day-to-day activities and not something that can be remembered when something goes wrong. In our experience, this can be achieved by linking responsibilities for OSH to:

1   Job descriptions.
2   Performance appraisal systems.
3   Performance and reward systems.

An example is given below.

OSH responsibility and performance management elements	
Item	**Description**
Job description insert (covers all managers and employees)	'To undertake your job responsibilities in accordance with organisation policies for OSH, and to draw the attention of your manager to any aspects that you consider do not conform to those policies.'
Appraisal element (linked to job description insert) – manager's example	'There is an active management of OSH within their area of responsibility, and an effective implementation and compliance to company policies.'
Employee example	'There is co-operation with, and compliance to, organisation policies for OSH.'
Performance criteria (linked to appraisal element) – manager's examples	Specific goals and 'measurements' will be included based on current status – primarily based on 'inputs' (work to be done), eg implementation of specific OSH procedures, rather than 'Outputs' (consequence measurements), eg reduction in costs of vehicle accidents or employee occupational health/accident claims.

Item	Description
	Examples:
	OSH is a prominent 'agenda' item at appropriate management and employee meetings
	OSH is always a factor that is taken into account within day-to-day business and operational processes
	The management of OSH is reviewed at regular intervals in accordance with the organisation's monitoring procedures
	For managers of employees who undertake work-related driving, the specific risks are proactively managed, including the managers' direct involvement in control measures, eg initial driver assessments and regular observations of drivers as part of their 'performance review'. The level of accidents and unrecoverable costs show a declining trend
	'Hazard control' activities are managed effectively, resulting actions are implemented and maintained, and the number of actions show a declining trend
	Employees all receive appropriate training and receive regular communications about OSH
	There is active co-operation with the organisation's OSH adviser
Employee examples	Specific goals and measurements for employees could include hazard-reporting activity, involvement in OSH activities, reduction in personal accidents – vehicle or work activity.
	Examples:
	Actively co-operates with organisation's policies for OSH
	Actively reports 'hazards' and other deficiencies in the arrangements for OSH
	Reads communications, attends briefings and training courses related to OSH
	Only undertakes work for which they feel competent and where training has been provided

Item	Description
	Works in accordance with OSH policies and does not put themselves or others at risk
	Adopts defensive driving practices, especially during any work-related driving activities
	Complies with the organisation's policies on work-related driving and the use of mobile telephones
	Reduces their personal creation of hazard-reportable incidents and their individual accidents

Communications

5.22 Communications for OSH should be a two-way process, and should be part of the organisation-wide communication processes for risk management. The organisation should implement processes that allow for and encourage communications on OSH between managers and employees. In addition, there is a need to provide for effective communications with external organisations, including regulatory authorities. The process must not be used as a substitute for formal training. Communications can be between an individual employee and their line manager, trained/authorised trainer or more experienced employee, or in a group environment. Promotional methods should also be used to advertise OSH, including particular developments, key messages, goals and performance/progress, statistical analysis, bulletins, etc, which can raise the profile of OSH and provide a reminder that OSH is a part of normal business processes, including risk management.

5.23 The principles behind effective communications and the role they can play in the management of risk, including OSH risks, are discussed in detail in **Chapter 6**, but some key points need to be highlighted:

- A formal assessment should be made to identify the communication needs of the organisation, and the following factors should be considered to aid process development:
  - Total numbers of employees and line managers within the scope of the OSH management system.
  - Geographical spread and organisation of site- and field-based employees.
  - Degree and level of hazards at the site and in the field.
  - Current communications processes and training arrangements.
  - Current promotional activities.

- Need for all employees to have regular access to promotional information.
- Need to include a record of the instruction, communication or promotional activity.

5.24 The outcome of this assessment should be discussed with managers and employees and decisions made about the communications processes needed by the organisation

- Communication methods – these should be selected carefully to fit the organisation's existing systems, so that OSH information is provided via the main organisational process, and not a separate system that adds to cost and is not likely to take place. Methods typically used are:
  - Management meetings – risk management and OSH should be a standard agenda item. Discussion items can include:
    - Management systems status and progress.
    - Creation and implementation of new initiatives.
    - Monitoring, audits and recommendations.
    - Recent accidents and product incidents.
    - Progress on actions from above.
    - New legislative requirements.
    - Enforcing authority inspections and audits.
    - General promotional initiatives.

5.25 The minutes/notes/action points should be recorded.

- Individual communications between a manager and an employee – these should cover general OSH aspects that are important to understand within the general work environment. Examples include a reminder about general safety rules, hazard reporting, maintaining machine guards and wearing PPE. It is also vital that employees are given appropriate information to ensure they can participate in the risk management and OSH systems. The key information required by employees falls within three segments:
  - Recruitment-based information – job description and job specification against which they are assessed and appointed.
  - Modifying and improving their existing skills and competencies, against which they are trained and improved.
  - Monitoring and feedback against which they are assessed on goals and given feedback on performance to date and future development plans.
- Group communications – a regular process should be established that effectively provides appropriate information to all employees via a system of workplace briefings. The frequency, length and content of these briefings should be varied as required, taking account of the availability of employees, the hazards/risk with the working environ-

ment and other factors. The process should be co-ordinated with communications on other items, eg business information, quality, production, financial, to maximise the benefits and minimise disruption to normal day-to-day activities.

- Information and promotional activities – these should be developed for each part of the organisation and specific responsibility for organising these activities should be allocated based on the organisational needs, coupled with adequate time and resources to make the process effective.

- Proportionate distribution – the vital principle of proportionate distribution, ie circulating information to the point at which it has to be acted upon in a manner that enables effective action. Too often, all documents are circulated to everyone who may possibly need to refer to it, without any thought given to proportionate distribution. Sometimes it is possible to detect a 'tick-in-the-box' mentality, eg 'Well we have sent it to them, now it is not our fault if they don't comply – and we will "beat them up" when they do not'. A process should be established that differentiates between different levels of management and different functions and only circulates information that is relevant. So the Chairman/CEO/MD will be notified of a change using an executive summary; whereas a line manager who has to implement a process will be given full details, plus supporting information and potentially an awareness/training session to ensure they understand the principles involved, their responsibilities and the required actions. Such a process is vital if managers are to be assisted to understand, accept and act upon their responsibilities.

- Current policies and procedures – designing, creating and publishing a full OSH 'manual' will neither ensure that anyone will read the content or act upon it. It is vital that effective communication systems are used to bring the information to the attention of those who have to act upon it. Transferring the content to an internal 'Intranet' will achieve the goal of making the information accessible to most within the organisation, but will still not ensure they know it is available, can actually find it, and, most importantly act upon it.

- New initiatives – these can arise for a number of reasons, including:
  - Changing business requirements, eg new products, new process, new machinery.
  - New regulatory requirements, eg work-related driving, asbestos in the workplace.
  - OSH management system requirements arising out of reporting and monitoring activities, eg risk assessments, accident reports, hazard control.

5.26 OSH professionals should initially evaluate whether there is a need for a new initiative, or a change in existing arrangements. The type of change will be discussed with the affected areas, and a plan will be created to ensure that all affected functions and people are consulted, prior to any change being implemented. The need for a new initiative will be reported to the organisation's employee consultative group – often called the 'OSH Committee'. A chart we designed for a global organisation is set out in Figure 5.2.

- Employee consultation – there is a legal requirement to establish effective processes for communicating with employees about the management of OSH. Although this does not extend to risk management, there are sufficient business benefits in ensuring that employees are fully informed and consulted about all aspects of organisational and operational risk. The actual arrangements will depend on each

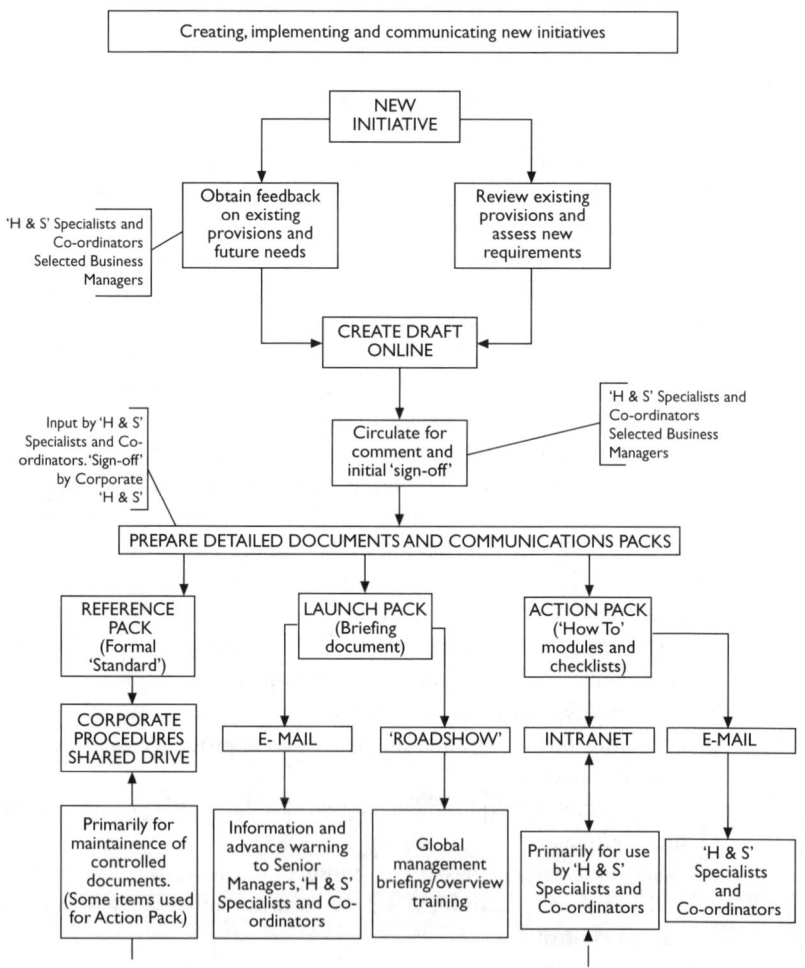

FIGURE 5.2

organisation, but the process implemented should be created with the intention of making it successful and not just a mechanism to satisfy regulations. The process should focus on management system matters at the tactical and strategic levels – not the operational levels.

- External compliance liaison – a process should be established to ensure that there are effective arrangements for liaising with and responding to visits and information requests from any body that has OSH responsibilities, eg Health & Safety Executive and local authority environmental health enforcement officers, fire brigade.
- External liaison – there should be a process for ensuring that information is available to external parties, eg customers and suppliers, to enable the organisation to respond to requests for OSH information, including copies of the 'policy' (see above).

Recruitment, selection, learning and training

Introduction

5.27 The organisation should have a process – which includes appropriate OSH aspects – to recruit, select and initially place new and transferred employees, a prior assessment of required physical capabilities and a medical to identify current health status where genuine occupational requirements can be identified. The process should also include reference to OSH policies, organisational culture and expectations and highlight key risks and control methods.

5.28 Lack of knowledge and skill is a significant underlying cause of OSH-related loss and it is vital that effective knowledge and skills training provides employees with the understanding and competence to work in a safe and efficient manner. Training is the glue that binds the elements of an activity into a safe and effective process and therefore effective training is vital to the successful management of the whole business, including the OSH management system.

Recruitment and selection

5.29 The recruitment and selection criteria for a job should be based on an assessment of the activities/tasks and job demands, required training, skills and knowledge, general human resources, legal and OSH aspects.

5.30 The OSH aspects should be defined using the following inputs:

- Details from the risk assessment and systematic survey processes related to physical requirements.
- A review of all occupations by the organisation's medical practitioner of particular requirements, based on a study of job tasks/activities, demands and potential hazards/risks.

- Health hazard analysis and monitoring.
- Particular job requirements eg work related driving, wearing respiratory protection equipment.

5.31 The OSH aspects should be used during recruitment and selection and to determine the content of pre-employment medicals.

PRE-EMPLOYMENT MEDICALS

5.32 All organisations should give serious consideration to undertaking pre-employment medical screening, followed up by full medicals where the initial screening identifies that further investigation is required.

5.33 Records should be kept of these examinations in a secure, controlled, confidential storage facility to which only the company medical practitioner has access.

REFERENCE AND QUALIFICATION CHECKS

5.34 Subject to legal requirements, the recruitment and selection process should include references from previous employers, qualifications, driving history and driving licence status, where appropriate.

Learning and training

POLICY

5.35 Organisations should establish a clear statement of policy that:

'No employee is expected to undertake an activity or task if they do not have the experience, the training, skills and equipment that create the competence to perform the task safely and efficiently.'

5.36 The organisation should also provide necessary resources to ensure that appropriate OSH training is included within business and employee training plans, to ensure that all managers and employees are aware of:

- The fundamentals of OSH regulations.
- Key OSH policies and a broad appreciation of the organisation's OSH management system and related business-risk management systems.
- Their responsibilities within the OSH management system, and related business risk management systems.
- The outcome of risk assessments, and the actual or potential hazards/risks arising from their work activities/tasks.
- Specific OSH policies and procedures that apply to their work activities.

TYPE AND LEVEL REVIEW PROCESS FOR AN INDIVIDUAL EMPLOYEE

5.37 The following aspects should be used to determine the required type and level of OSH learning and training for an individual employee:

- Job responsibilities, personal and team goals and objectives.
- Work activities and actual/potential hazards/risks arising from risk-assessment processes, and system monitoring.
- Individual development plans.

5.38 The outcome of the review will be used to link to the organisation's core training programmes and to build the required learning and training into the personal/departmental learning and training plans, which will form part of the overall organisational process. An example of an overall process is set out in Figure 5.3.

5.39 Specific elements that need to be considered are:

- **Induction** – the organisation should have a comprehensive induction process for new and transferred employees that provides appropriate information during the early stages of their appointment. This should include:
  - General information about the organisation's history, culture, structure, products/services, etc.
  - General information about OSH policy, quality and financial systems, etc.
  - More detailed information about rules related to OSH hazards, risks and control methods.
  - Detailed information about general and specific responsibilities and department functions.

5.40 Depending on the complexity of the organisation, consideration should be given to using a separate induction process for managers that can then act as an introduction to more formal training to be given at a later date.

5.41 The induction process can be divided into segments:

- General induction on the first day of appointment.
- Departmental induction during the second day of appointment.

5.42 The process should be documented, responsibilities allocated, check-lists created and used and records kept. The line manager should review the induction process with the new/transferred employee to ensure the information has been retained.

- **Management and leadership training** – All managers with responsibilities for OSH must receive initial training to ensure that they understand the organisation's management system for OSH and their responsibilities within it. Additional training may need to be given on specific OSH business processes.
- **Skills and competencies** – The organisation's recruitment and selection process should have identified the current skills and competences of the new employee, using specific assessments and/or by requiring evidence of qualifications. A probationary period may be required to

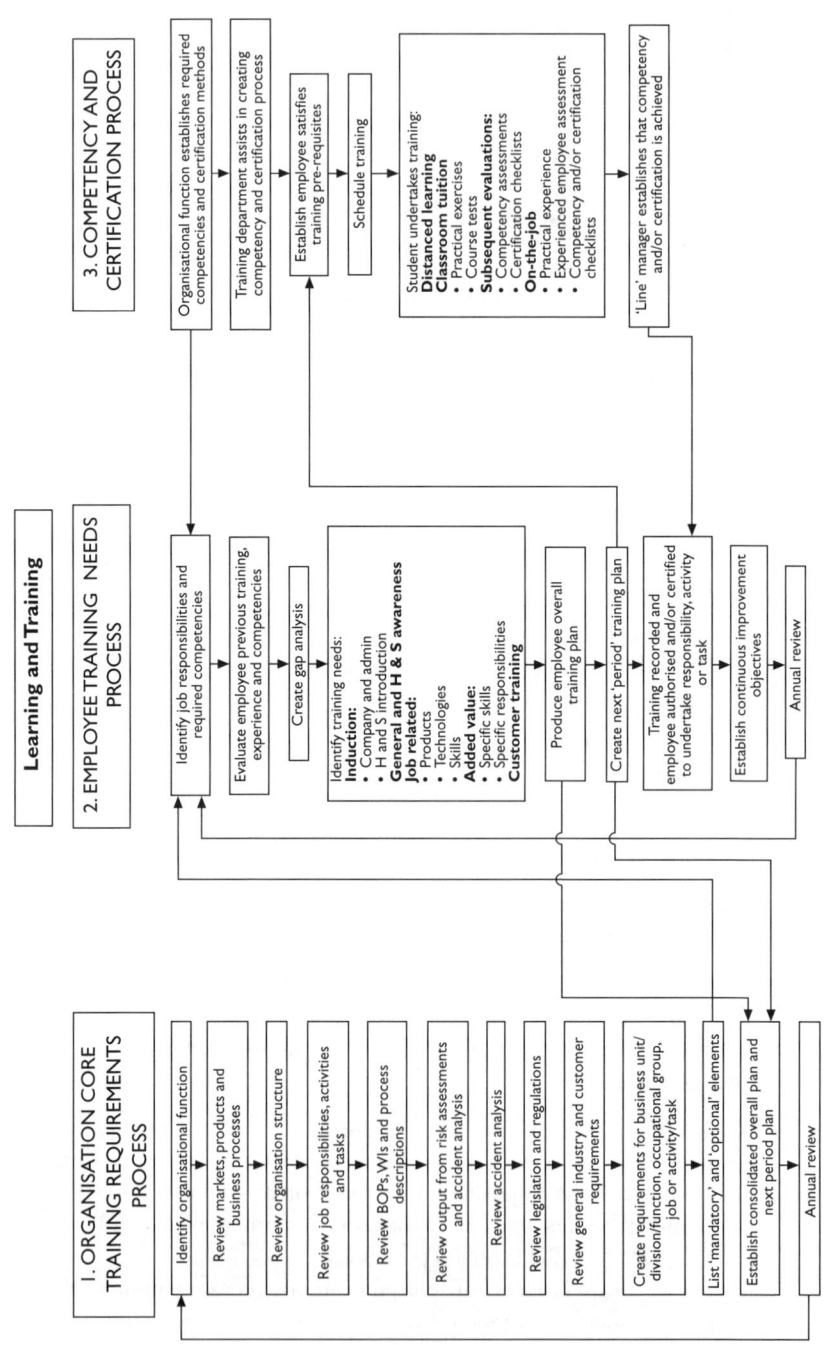

Learning and Training

1. ORGANISATION CORE TRAINING REQUIREMENTS PROCESS

- Identify organisational function
- Review markets, products and business processes
- Review organisation structure
- Review job responsibilities, activities and tasks
- Review BOPs, WIs and process descriptions
- Review output from risk assessments and accident analysis
- Review accident analysis
- Review legislation and regulations
- Review general industry and customer requirements
- Create requirements for business unit/division/function, occupational group, job or activity/task
- List 'mandatory' and 'optional' elements
- Establish consolidated overall plan and next period plan
- Annual review

2. EMPLOYEE TRAINING NEEDS PROCESS

- Identify job responsibilities and required competencies
- Evaluate employee previous training, experience and competencies
- Create gap analysis
- Identify training needs:
  Induction:
  - Company and admin
  - H and S introduction
  General and H & S awareness
  Job related:
  - Products
  - Technologies
  - Skills
  Added value:
  - Specific skills
  - Specific responsibilities
  Customer training
- Produce employee overall training plan
- Create next 'period' training plan
- Training recorded and employee authorised and/or certified to undertake responsibility, activity or task
- Establish continuous improvement objectives
- Annual review

3. COMPETENCY AND CERTIFICATION PROCESS

- Organisational function establishes required competencies and certification methods
- Training department assists in creating competency and certification process
- Establish employee satisfies training pre-requisites
- Schedule training
- Student undertakes training:
  Distanced learning
  Classroom tuition
  - Practical exercises
  - Course tests
  Subsequent evaluations:
  - Competency assessments
  - Certification checklists
  On-the-job
  - Practical experience
  - Experienced employee assessment
  - Competency and/or certification checklists
- 'Line' manager establishes that competency and/or certification is achieved

FIGURE 5.3

check the new employees' actual type and level of skills and competencies. Under no circumstances should proof of a skill or competency be assumed based on historical evidence, especially from another organisation. Where new skills are required, specific training must be provided before an employee is allowed to undertake a particular task.

5.43 For example, even where a new employee produces written evidence that they are trained on a particular type of forklift truck, it is vital to ensure that the training was for the particular type of forklift truck to be operated, and most importantly that an assessment of their skills in operating the particular forklift truck is undertaken in a controlled environment, and not in an operational environment. Similarly, where a new employee is required to wear respiratory protective equipment (RPE), the recruitment and selection process should have determined whether they were physically capable of wearing such equipment, eg did they suffer from any respiratory condition that would be exacerbated by wearing such equipment? In addition, prior to wearing such equipment for the first time in an operational environment, the new employee should be trained on the fitment and usage of the equipment to ensure an adequate level of skill and competency.

Risk assessments

Purpose and application

5.44 The purpose of risk assessment is to:

- Undertake an identification of existing and potential hazards.
- Evaluate hazards for their potential to cause harm to employees and others affected by organisational activities.
- Create action plans to implement improvements in controls that will reduce the risks to tolerable levels.
- Review the situation on a regular basis based on the level of risk, and the type of controls in place.

5.45 Interestingly, and often with disturbing and tragic results, the last two items in the above list are not completed. Many organisations believe that risk assessment is about identifying the hazards, completing the assessment and then filing away the results. Indeed, some go even further by only assessing the most obvious hazards, and certainly not those that are called 'input' and 'output' hazards.

5.46 Figure 5.4 demonstrates the relationship between input risks, activity risks and output risks. In addition, the figure also demonstrates that, when looking at risks, the environment within which these processes are taking place is also important. This concept will be discussed in more detail in **Chapter 7**.

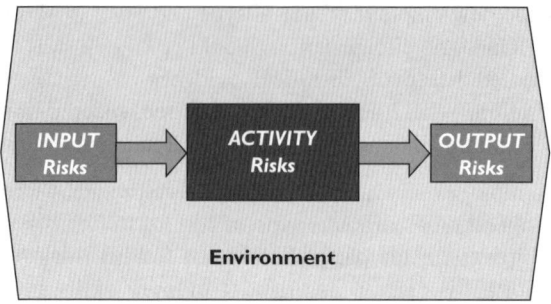

FIGURE 5.4

5.47 Risk assessment processes should be designed to explain how the organisation is going to apply regulations/organisational standards, and reduce the risk to a tolerable level by the introduction of policies, procedures and other control methods that are appropriate to the organisation and the level of risk in each situation. Risk assessment policies and procedures are not designed to just record the results of the assessment and be filed.

Categories of risk assessment

5.48 Risk assessments are required by various pieces of legislation – some general and some specific. The following are the main risk assessment categories that organisations need to consider:

- General (MHSWR – Management of Health and Safety at Work Regulations 1999).
- Display-screen equipment (DSE – Health and Safety (Display Screen Equipment) Regulations 1992).
- Provision and use of work equipment (PUWER – Provision and Use of Work Equipment Regulations 1998), including complex machines.
- Manual handling.
- Ergonomics.
- Control of Substance Hazardous to Health (COSHH – Control of Substance Hazardous to Health Regulations 2002).
- Fire and emergencies.
- Asbestos.
- Lead.
- Noise.
- Lone working.
- Home working.
- Working at height.
- Confined space working.
- Work-related driving and mobile telephone usage.

- Young persons.
- Pregnant and nursing mothers.
- Stress.
- External working, eg sales and service external activities.
- Personal protective equipment (PPE – Personal Protective Equipment Regulations 2002).

5.49   Each category may potentially require its own process, which should be integrated with existing assessment processes for other management systems, eg business risk management and environmental management. Each process should be proportionate to the general risk profile of the organisation. For example, an organisation in the semi-conductor industry will require more detailed and technical processes than, say, a clothing retailer. However, the clothing retailer will have risks that the semi-conductor manufacturer will not.

Process

5.50   It is vital that a structured process is used to manage the risk assessment process. An uncoordinated and haphazard approach will not identify the significant hazards/risks, and is more likely to encourage a reactive approach, following something going wrong.

5.51   In Figure 5.5 we show how a general risk process is used to determine whether further actions are needed to enhance risk controls and whether a detailed risk assessment is needed.

5.52   We recommend the following steps be implemented – even where an organisation believes that it has an effective current process.

1   Divide the 'site' in manageable areas, ideally using an 'accountability map', allocating every area to a line manager. Remember to cover all 'public/general' areas, eg stairways, corridors.

2   Select a department/section/job/task (as appropriate) and allocate the risk assessment task to a team of knowledgeable people who are advised by a trained risk assessor.

3   Identify the activities and hazards/risks arising within the selected area, including hazards within the 'input' and 'output' activities.

4   Use a general risk assessment process to see whether existing controls have reduced the risks to tolerable levels.

5   Using information obtained, identify any need for additional controls.

6   If more information is required, or the level of risk needs further evaluation in an area of risk where the assessor has not been trained to an 'advanced level' or does not possess the required competency, eg, specific noise risk assessment, then the risk assessor should discuss the situation with the organisation's OSH professional to decide on the next step(s).

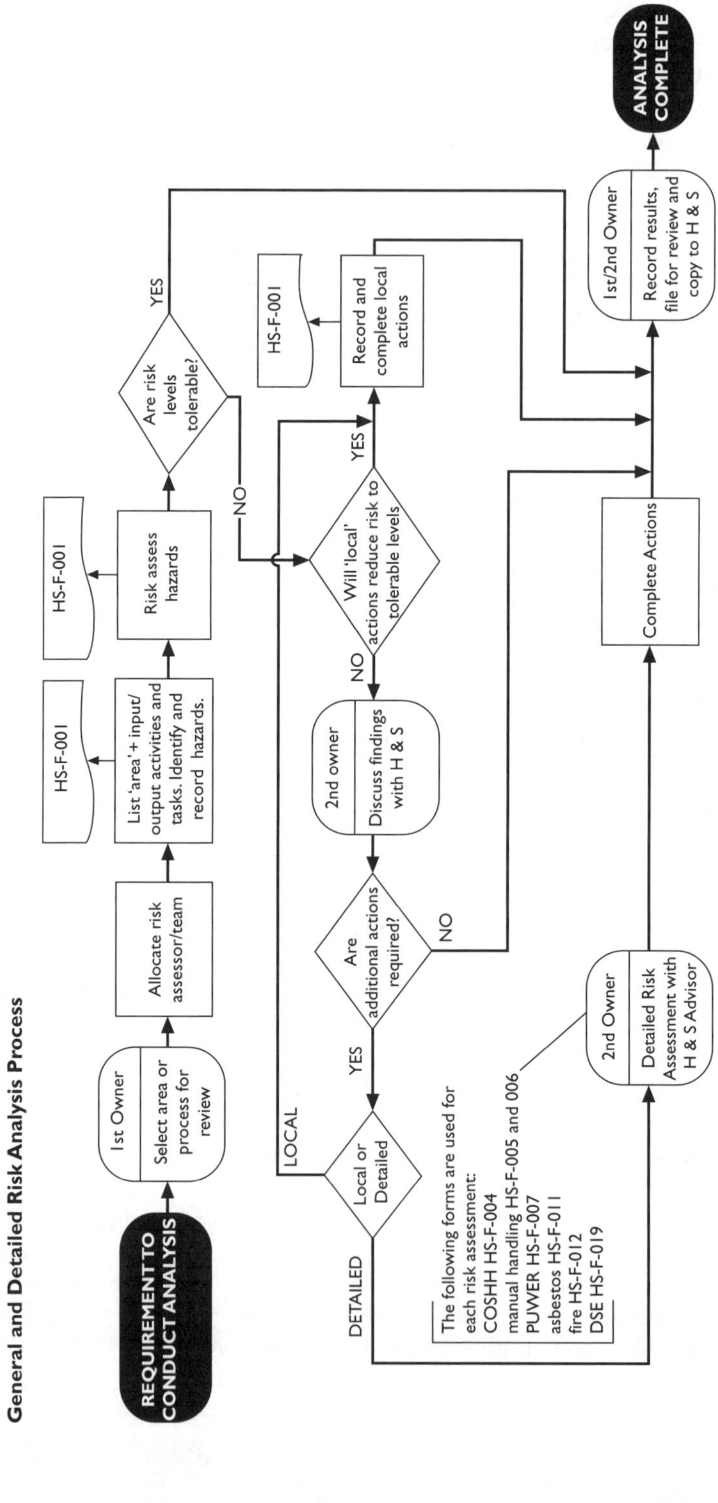
General and Detailed Risk Analysis Process
REQUIREMENT TO CONDUCT ANALYSIS
1st Owner
Select area or process for review
Allocate risk assessor/team
List 'area' + input/ output activities and tasks. Identify and record hazards.
HS-F-001
Risk assess hazards
HS-F-001
Are risk levels tolerable?
YES
NO
Will 'local' actions reduce risk to tolerable levels
YES
Record and complete local actions
HS-F-001
NO
2nd owner
Discuss findings with H & S
Are additional actions required?
NO
YES
Local or Detailed
LOCAL
DETAILED
The following forms are used for each risk assessment:
COSHH HS-F-004
manual handling HS-F-005 and 006
PUWER HS-F-007
asbestos HS-F-011
fire HS-F-012
DSE HS-F-019
2nd Owner
Detailed Risk Assessment with H & S Advisor
Complete Actions
1st/2nd Owner
Record results, file for review and copy to H & S
ANALYSIS COMPLETE

FIGURE 5.5

7   Potential next steps are: (a) decide that the controls are acceptable, and take no further action; (b) undertake a detailed risk assessment of a particular type, eg manual handling, COSHH, noise, using either an in-house or external specialist.

8   Determine the required actions to implement new controls, allocate responsibility for each action and set timescales

An example actions schedule is set out in Figure 5.6.

Risk level		Action timescale
Significant	Very high	Immediate
	High	4 hours 1 day
	Moderate	1 week 2 weeks
	Low	3 weeks 1 month
	Very low	2 months 3 months
Insignificant		N/A

FIGURE 5.6

9   Implement and monitor the results, and review the risk assessment on a periodic basis, as determined by its original level of risk and existing control method, eg a risk that requires a permit to work to control the exposure will need to be reviewed more frequently than a risk that can be controlled by a general OSH rule.

An example review schedule is set out in Figure 5.7.

Risk rating		Control	Review period
Significant	Very high	Permit to work	1 year
	High		2 years
	Moderate	Written instructions	3 Years
	Low		4 years
	Very low	Verbal instruction by competent person	5 years
Insignificant		N/A	N/A

FIGURE 5.7

Rules and procedures

Introduction

5.53   The term 'Rules' is used to collectively describe the different types of rules outlined below in the hierarchy of rules.

5.54   It is often necessary to use documentation in various formats to communicate what is required in a particular set of circumstances, supported by appropriate learning, training and communication. It is important to use the correct type/level of Rule that is proportionate to the type/level of control required, as an 'over-the-top' Rule can be as ineffective as no rule at all, or a badly drafted/implemented rule.

Process

5.55   The type/level of Rule that is required to ensure effective control is influenced by the outcome of the risk assessment, especially any detailed risk assessments. When all reasonably practicable elimination, substitution and physical controls have been implemented, and further controls are still required, then consideration should be given to supporting Rules, using this hierarchy of Rules:

- Permit to work.
- Chemical/mechanical/electrical isolation.
- Safe system of work (SSW).
- Method statements and work instructions (WI).
- General OSH management system procedures.
- Signs and labels.
- General safety rules.
- Verbal instruction from a line manager.

Integration, not duplication

5.56   Any Rules should be designed to dovetail with existing management systems for quality, environment, finance, etc, and should not be created as a completely new set of Rules, unless absolutely necessary. In particular, organisations may have included the OSH management system as part of the scope of an ISO 9000 quality system or an ISO 14001 environmental management system.

5.57   If this is the case, then the documentation used will have to follow the primary system – normally ISO 9000 where it exists. This may require OSH management system Rules to be described in a different format, with more focus on flowcharts/diagrams rather than long written documents. This has implications for communications processes, discussed earlier. For example, where existing methods of work cover an operational activity, any additional requirements – related to business, operational, OSH or environmental risks – should be added to the existing documents, rather than separate documents being created. In our experience, the Rule that relates to an operational activity is most frequently used or referred to, so it is good practice to use that as the base document. Entering additional requirements as part of an operational activity 'rule', rather than a separate 'rule' that can easily be forgotten or misunderstood as to its application, is the safest, most effective and efficient process. Additionally, training and communication can be undertaken using 'one' document so that managers and employees see the management of the activity as a total organisational process.

5.58 Sometimes specific Rules will be needed to support the activity, eg hot work permit, but the standard operational 'rule' should include the requirements for the use of a hot work permit, leaving the only additional document to be the actual hot work form itself.

5.59 Where the operational activity is not 'standardised' then a process must be used to create a 'rule' that is specific for the activity. A process that requires a safe system of work can look like that set out in Figure 5.8.

5.60 Using a formal consistent process to create specific Rules will ensure that all factors are taken into account and will build up a bank of knowledge within the organisation of how certain activities should be managed. This bank of knowledge can be used to influence work changes (see below) and can be used during training programmes. Additionally, the organisation will create a confidence (not complacency) about how activities can be managed that will prevent ad hoc decisions being taken when urgent situations arise.

5.61 OSH professionals can point to the formal process for creating a specific 'rule', but can also use examples and experience of applying the process to other circumstances. This reduces the potential for managers to use 'make-it-up-as-you-go-along' or 're-invent the wheel' approaches every time something that is not already covered by formal processes needs to be managed. How often have OSH professionals heard comments like 'You're right, that has happened before, but we can't remember what we did or how it was resolved, so we did what we thought was right at the time, with unfortunate consequences, but it will not happen again, because we always learn from our mistakes'?

5.62 Our experience is that organisations who decide to apply consistent processes for creating and using Rules and adopt an approach of organisational learning will be much more effective in managing risks. The benefits are considerable and also help to reduce any tendency for a blame culture to be created.

5.63 The following points indicate that a learning process is in place: adopting a culture of learning from previous situations; being prepared to

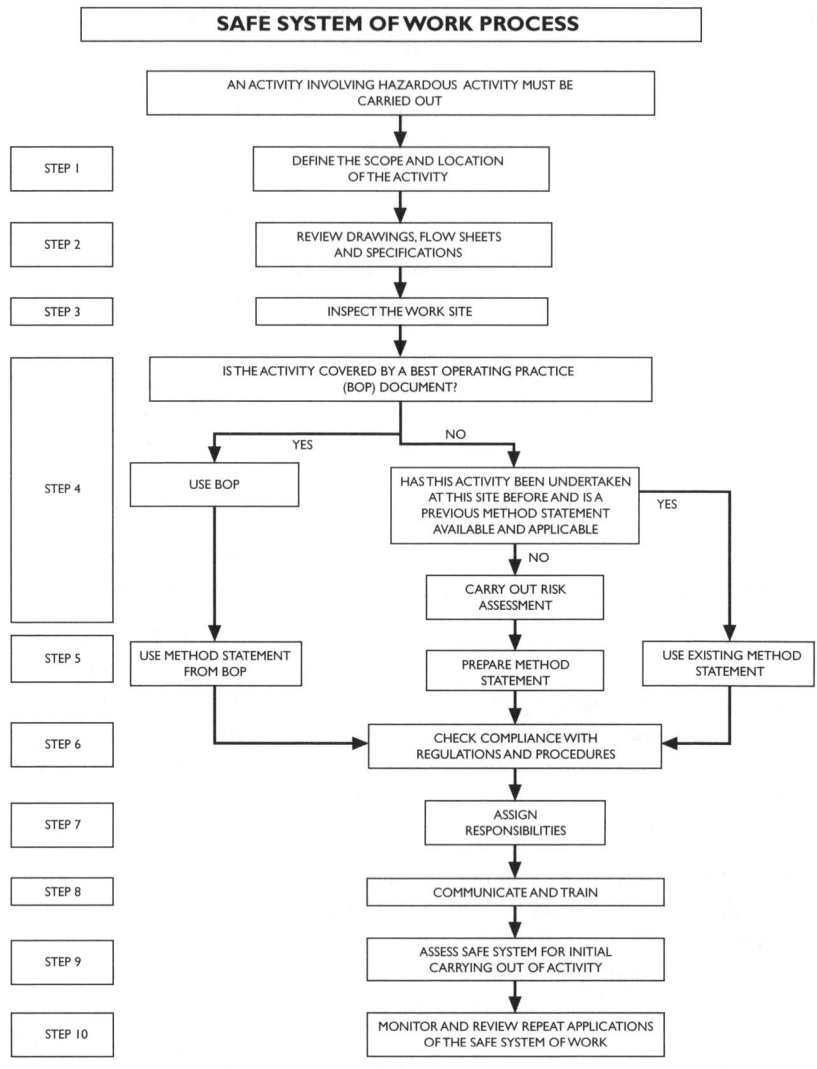

FIGURE 5.8

face up to the reality of 'what went wrong' (not blame allocation); adopting systems to capture what went wrong, why and the actions to prevent a reoccurrence; establishing a formal process for future 'rule' creation; ensuring training and communication for those Rules is effective; and using existing Rules as reference points. All of the above will reduce the risk of the same set of circumstances happening again. OSH professionals dread these words from someone involved in the accident: 'You know that has happened before. Did we not learn from that and did we not decide to implement a new system to stop it happening again – but it did!'

Health and hygiene

Introduction

5.64 A major cause of workplace accidents is the lack of occupational health provision, resulting in work-related diseases and illnesses. Over the last few years the trend for 'injuries' is decreasing, whereas the trend for 'ill-health' is increasing. This is particularly significant in the areas of asbestos-related diseases, stress and occupational asthma. Tackling these issues on a piecemeal basis or reactively will not manage the risks. OSH professionals should therefore ensure that there is a co-ordinated and consistent health and hygiene system that is fully integrated with other organisational processes, particularly Human Resources processes.

Policy

5.65 An organisation should consider the publication of a more detailed additional section to their OSH policy. Where required, the creation of such a section will ensure that managers and employees appreciate the significance of health and hygiene hazards/risks and the resources being made available by the organisation to manage those risks. Although some policy items will apply to all organisations, others will depend on the size and geographical spread of the organisation and its risk exposures.

5.66 An example of a more detailed statement is as follows:

- To ensure, as far as is reasonably practicable, that employees' health problems are not caused by the workplace environment or work practices and that specific groups of employees are medically supervised as may be required by regulation and/or risk assessment.
- To ensure that the organisation's risk assessment processes include consideration of health hazards/risks.
- To ensure that methods of work include requirements that manage exposure to health and hygiene risks.
- To ensure that training and communication arrangements include advice and instruction on controlling exposure to health and hygiene risks.
- To ensure that access to occupational health records is strictly controlled and restricted to the organisation's occupational health advisers. Such records will not be available to others within the organisation without the written consent of the individual. All records containing medical surveillance information will be kept in line with current legal requirements (40 years), after which time they will be destroyed. The arrangements also need to conform to Data Protection regulations, with particular reference to recent consultations/regulations regarding maintaining medical records under data protection.
- To make available to all employees an occupational health service, staffed by a qualified occupational health adviser or a medical

practitioner with occupational health experience, which is confidential, independent and impartial.

- To provide a treatment service for occupational injuries and illnesses including the supervision of emergency care and first-aid requirements for the general needs of the organisation.
- To ensure that there are adequate arrangements for the co-ordination of first-aid personnel in the event of a major emergency (see below).
- To constructively and sympathetically deal with employee problems related to the abuse of alcohol or other mood-altering drugs, including those described by a medical practitioner.
- To promote a healthy lifestyle amongst employees.

Processes

5.67   Specific management processes are required to make the policy effective. These should be integrated with other management systems, particularly those of Human Resources. These can include:

- Job and person specifications, eg any special requirements, linked to recruitment and selection processes, and specific emergency procedures.
- Pre-employment screening and medicals, linked into recruitment and selection processes, and the selection for specific emergency response activities.
- Special medical assessments, eg driving forklift trucks, linked to training and development processes.
- General occupational health service:
  - Medical records storage, restricted access and Data Protection.
  - Health promotions.
  - Personal counselling (stress, smoking, alcohol).
  - Return to work interviews.
  - Sickness absence monitoring.
- Risk assessments, especially COSHH (linked to environmental management processes, including spill/release emergency procedures), manual handling, DSE and confined space entry.
- Business process and operational changes linked to the management of contractors, and the purchase of new materials and equipment.
- Preventive maintenance, eg testing and service of local exhaust ventilation (LEV).
- Ongoing workplace monitoring, eg atmospheric sampling, linked to preventive maintenance.
- Health surveillance, eg periodic tests of lung function, blood and urine.
- Personal hygiene support, eg control of exposure to substances in relation to PPE fitment, linked to training and workplace communication processes.
- Environmental assessments linked to health surveillance.

A typical integrated process for managing health and hygiene is set out below:

U.K. Occupational Health Service
Process Owner: Health & Safety Advisor

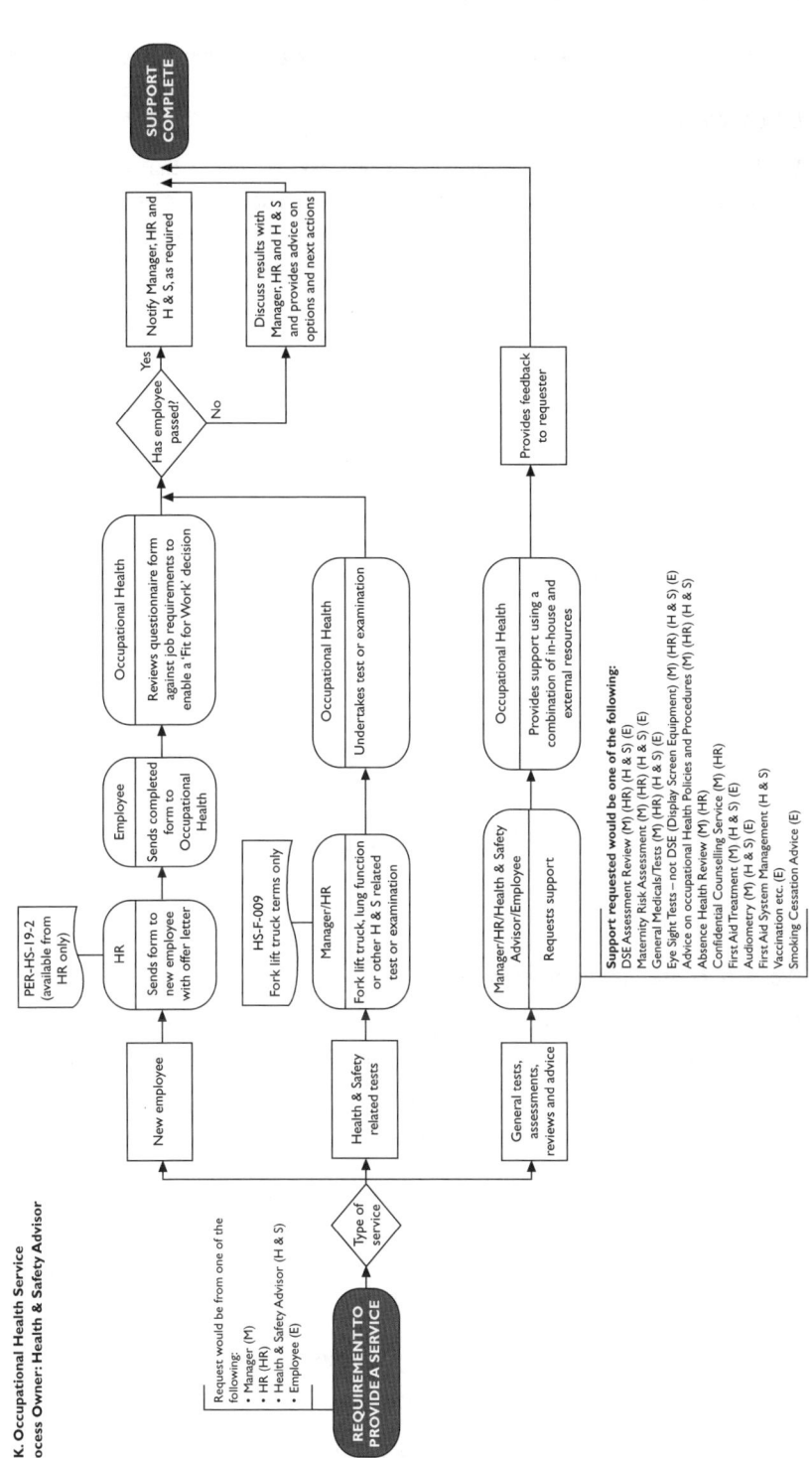

Request would be from one of the following:
- Manager (M)
- HR (HR)
- Health & Safety Advisor (H & S)
- Employee (E)

REQUIREMENT TO PROVIDE A SERVICE

Type of service

New employee

PER-HS-19-2 (available from HR only)

HR
Sends form to new employee with offer letter

Employee
Sends completed form to Occupational Health

Occupational Health
Reviews questionnaire form against job requirements to enable a 'Fit for Work' decision

Has employee passed?

Yes — Notify Manager, HR and H & S, as required

No — Discuss results with Manager, HR and H & S and provides advice on options and next actions

SUPPORT COMPLETE

Health & Safety related tests

HS-F-009
Fork lift truck terms only

Manager/HR
Fork lift truck, lung function or other H & S related test or examination

Occupational Health
Undertakes test or examination

General tests, assessments, reviews and advice

Manager/HR/Health & Safety Advisor/Employee
Requests support

Occupational Health
Provides support using a combination of in-house and external resources

Provides feedback to requester

Support requested would be one of the following:
DSE Assessment Review (M) (HR) (H & S) (E)
Maternity Risk Assessment (M) (HR) (H & S) (E)
General Medicals/Tests (M) (HR) (H & S) (E)
Eye Sight Tests – not DSE (Display Screen Equipment) (M) (HR) (H & S) (E)
Advice on occupational Health Policies and Procedures (M) (HR) (H & S)
Absence Health Review (M) (HR)
Confidential Counselling Service (M) (HR)
First Aid Treatment (M) (H & S) (E)
Audiometry (M) (H & S) (E)
First Aid System Management (H & S)
Vaccination etc. (E)
Smoking Cessation Advice (E)

Property conservation, business continuity and emergencies

Introduction

5.68   The organisation should have a process to ensure that the structure and fabric of the organisation's property is constructed, arranged and maintained to minimise hazards, reduce losses and remove risks to people. The organisation should also ensure that premises are well protected by fire detection and suppression systems, based on the level of risk and the effects of a loss.

5.69   Business continuity should be considered during decisions on new premises and new activities, as increasing internal interdependencies and external dependencies are crucial to maintaining the ability of the organisation to continue to undertake its activities, especially in the event of an emergency.

5.70   Equally the organisation should have in place an emergency response process that is designed to react to previously identified emergencies, so that the organisation's employees and others involved in or affected by its activities can be safeguarded.

5.71   OSH professionals should ensure that they are actively involved in all three processes to ensure that all relevant OSH risks are taken into account and that the organisation integrates these processes with the OSH management system. The OSH professional can play a key role in ensuring that all three processes take account of the outcomes of each other to create an overall plan for the organisation. In that way any changes in one area can be related to the other areas.

Property conservation process

5.72   It is important that an organisation understands its level of risk and the status of existing controls. Consequently, the following should be undertaken:

- Carry out a survey to identify the current risks and level of protection.
- Evaluate that status against prescribed standards that apply where the premises are located or by establishing some organisational standards that are applied throughout the organisation. This could be the application of specific standards, eg the US, UK or Australian standards, to the whole organisation or certain geographical regions, or the organisation creates its own standards which are at a level equal to or above any 'local' requirements.
- Consider the business continuity aspects.
- Submit recommendations for action.
- Evaluate the recommendations and prepare an action plan in conjunction with appropriate specialists and insurance advisers.

5.73 The organisation should consider any proposed changes to the following items to determine whether changes are required to the types or levels of protection being used:

- Usage of the premises.
- Production processes and equipment.
- Chemicals/materials used and stored.
- Value of inventory.
- Increased dependency linked to business continuity.

CONTROLS

5.74 Controls to reduce or eliminate risks to people and property can be physical or process related. There should be controls to:

- Maintain management and employee commitment to emergency arrangements.
- Install automatic fire detection, suppression and monitoring.
- Ensure good housekeeping and no-smoking policy.
- Ensure good security arrangements.
- Segregate and protect special hazards.
- Remove/segregate hazardous waste.
- Maintain comprehensive planned preventive maintenance (PPM) of fire equipment.
- Only use buildings that are designed for the activities to take place within the premises.
- Maintain good emergency preparedness.

5.75 The above process-related controls should form part of the organisation's emergency policy and plan. They should also be considered during the business continuity and emergencies processes.

Business continuity

5.76 The organisation should have an ongoing process that enables it to be fully prepared at all times to contain interruption to its activities, manage the consequences of emergencies and disasters, activate survival plans and ensure recovery to the integrity of the organisation.

5.77 The process should have the following steps:

- Undertake detailed business impact analysis, with particular reference to special activities, vital sources of supply, special premises, people and processes that cannot be replicated.
- Establish recovery objectives.
- Identify key facilities and resources.
- Identify and assess risks.

- Establish plans and arrangements to ensure the organisation's ability to achieve its recovery objectives.
- Undertake tests and practices of the arrangements.

5.78 Those undertaking the business continuity process should consider the outputs from the property conservation process, as these will indicate the risks to particular premises. In addition, consideration should be given to the ability of the organisation to respond to any emergency and the resources required to ensure an effective and safe response. These may well have been identified by risk assessment of the organisation's exposure to serious and imminent dangers (MHSWR 1999).

Emergencies

INTRODUCTION

5.79 This business process should be designed to prevent any unnecessary risk to employees, visitors, contractors and any other persons that may be on the organisation's premises at the time of a fire, fire evacuation, bomb threat or other serious or imminent danger, or, if external to the organisation, have the potential to be affected by an incident on the organisation's premises.

PROCESS

5.80 The process should have the following sections:

- **Emergency policy** – the overall policy, eg that the protection of people and not property is the priority; that only trained persons must tackle an emergency, all persons on the premises are to evacuate where instructed to do so.
- **Responsibilities** – define general and specific responsibilities for all those included in this process, eg overall co-ordinator, area co-ordinators, emergency response teams, line managers, employees.
- **Identifying requirements** – a systematic survey/analysis should be used to identify and evaluate emergency response needs of the organisation/site and any external implications, linked to fire and emergency risk assessments.
- **Developing the system** – use the results of the survey/analysis to develop a comprehensive emergency plan for general emergencies, eg fire evacuation policy, and specific requirements, eg fire evacuation plans and emergency response arrangements.
- **Internal plans** – using the overall emergency plan as a base, create a specific plan for each area of the organisation's operations for possible emergency situations, eg bomb threat, hazardous material spill clean-up, chemical release warning/notification, computer room evacuation/release of suppression system, protection of vital records, inter-

ruption of the telephone system, lift failure, storm, flood, utility services failure.

- **Plan communications** – a variety of methods should be used to ensure that necessary information is provided to all concerned, eg managers, employees, visitors, contractors, surrounding areas (see above).

- **Contacts** – internal and external emergency contact numbers should be listed, provided to those with responsibilities under this process and prominently displayed at key locations. The numbers should be formally checked at least annually.

- **Emergency communications system** – backup systems should be created and tested at least annually.

- **External liaison** – a process should be established to communicate information to external bodies about potential on-site hazards that could have implications for on-site activities of the emergency services and which have the potential to affect off-site areas (see above).

- **Media and next-of-kin liaison** – to ensure that accurate and timely information is provided to the media and all those concerned with an emergency and its effects, eg relatives of those potentially affected. Human Resources must be involved so that employees' next-of-kin contact details are available.

- **Training needs** – a link should be established with the core training needs analysis process to ensure all those with responsibilities under this section receive the appropriate training (see above).

- **Tests/practices** – the minimum frequency for testing monitoring, warning and protection systems, and holding system practices/simulations, should be determined by local legislation or organisational standards.

- **Plan reviews** – a formal review of each test/practice/simulation to objectively evaluate the results. The emergency services or other external bodies involved in the test/practice/simulation must be involved in any post-event review (a legal requirement under MHSWR 1999).

- **External Plans** – for employees based on a customer site or visiting customer sites, there must be adequate arrangements for their safety in the event of an emergency situation developing.

Hazard control

Introduction

5.81 In any workplace, if conditions deteriorate, the risk of personal and property damage accidents increase. The earlier that deviations from accepted standards are identified, the more effective the control of risk becomes. A proactive approach to identifying, assessing and eliminating hazards should therefore be adopted. The overall objective is to reduce the number, frequency and severity of accidents and losses, by adopting proactive approaches to identifying and eliminating hazards created by substandard conditions and/or substandard practices.

Policy

5.82   Typical policies for hazard control are:

- A work culture that encourages all employees to adopt a critical, OSH orientated approach to their own actions, the actions (or inaction) of others and to workplace activities in general. This can include providing training and communication to managers and employees on the legal, moral and business reasons for hazard control.
- Reporting by employees of any sub-standard condition or practice that they observe via a dedicated process. The term 'sub-standard condition' relates to the physical condition and environment of the workplace, machinery, tooling, personal protective equipment, etc. The term 'sub-standard action' relates to the below-standard or non-compliant actions of an employee, contractor or visitor. This is also a requirement under MHSWR 1999.
- A programme of scheduled general inspections by managers and OSH representatives, by type of workplace.
- A programme of specific inspections, eg forklift trucks, local exhaust ventilation.
- A means of systematically recording and rapidly following up reported 'hazards' from all OSH risk management system activities, including internally generated 'hazards' and externally reported actions, eg from enforcement officers.
- A process for statistically analysing hazard data to identify the need for additional hazard controls.

5.83   It is essential that all hazard control activities are co-ordinated as one complete process, so that the outputs from one element become the inputs for another. For example, the details from employee hazard reports should be considered by the person undertaking the planned inspection activity to make sure that additional items are inspected where necessary. Similarly, a manager undertaking a tour should review the results of the planned inspections to identify where the tour should focus.

5.84   Figure 5.9 shows the interlinked nature of some hazard control activities, and provides advice on typical frequencies for the activities

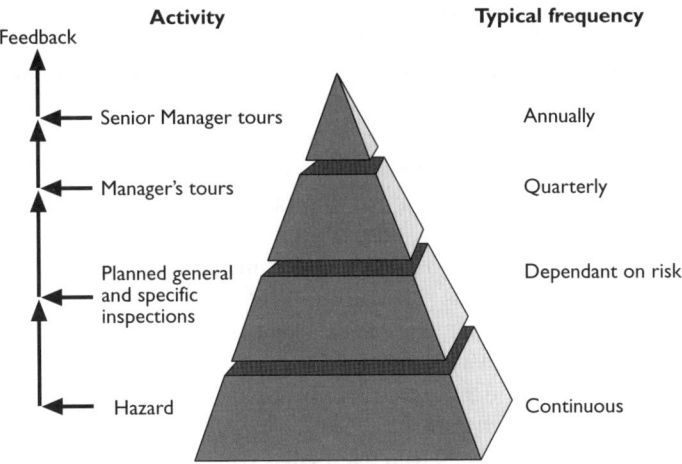

FIGURE 5.9

Process

5.85 A systematic process is required to ensure that all areas and activities are covered by the hazard control system. The process steps are as follows:

- Identify the sources required to ensure a comprehensive capture of relevant information. For example:
  - Employee hazard reports relating to sub-standard conditions and actions.
  - Maintenance requests.
  - Unscheduled maintenance (outside routine preventive maintenance activities).
  - General and specific site inspections/tours.
  - Emergency preparedness activities/practices.
  - Statutory and other 'regulatory' inspections.
  - Questions/points raised at OSH meetings/training.
  - Risk-assessment process.
  - Accident/incident reports.
  - OSH committee decisions.
- Divide the 'site' into manageable areas that are allocated to a specific manager. Depending on the size and hazard/risk profile of the area, consider allocating a dedicated person to the role of hazard control co-ordinator.
- Decide and list the specific hazard control tools that are needed for the area and develop a process to ensure that the hazard control activities are carried out, eg allocation of responsibilities for specific activities.
- Use organisational templates as a guide to generate area-specific checklists for local application, eg planned inspections. In our

experience, if area managers, together with the relevant supervisors/team leaders and safety representatives are encouraged to create their own lists, then they have a much greater ownership of the process, and are much more likely to undertake hazard control activities. In addition there is a need to determine the frequency of certain activities, eg planned inspections, manager's tours.

- Communicate or provide training (as required) to those with hazard control responsibilities, ensure awareness of hazards and their consequences for managing risks, which includes health hazards as well as safety hazards.
- Create an actions log that can be used to record the following:
  - Unique reference number.
  - Location of item.
  - Description of item.
  - Date occurred and date reported.
  - Name and information source.
  - Hazard classification and resolution timescales (see Figure 5.6).
  - Action details and comments.
  - Responsible person and target date.
  - Date actually completed.
  - Committee check ('sign-off' by local consultative group that hazard has been resolved).
- Use the actions log to analyse the data to identify trends that can highlight where OSH management system improvements are required.

An example of a process for an employee hazard report is set out in Figure 5.10 on page 153.

Personal protective equipment

Introduction

5.86   OSH professionals should ensure that the provision of personal protective equipment (PPE) is only a back-up system of control against a defined risk, and only used as a secondary control option.

5.87   The provision of PPE should not be approached, as is so often the case, as a reactive haphazard process that sees PPE as an adequate and, often, *only* control measure for the risk in question.

5.88   Engineering controls and safe systems of work should always be considered and implemented before PPE is actively considered. Management should look at the 'life-time' cost of the issue of PPE in any particular situation. It is often the case that the apparently 'expensive' option of implementing a more substantial control method, eg fixed barriers, will in fact be a less expensive option than, say, the issue of PPE for ten years to ten people.

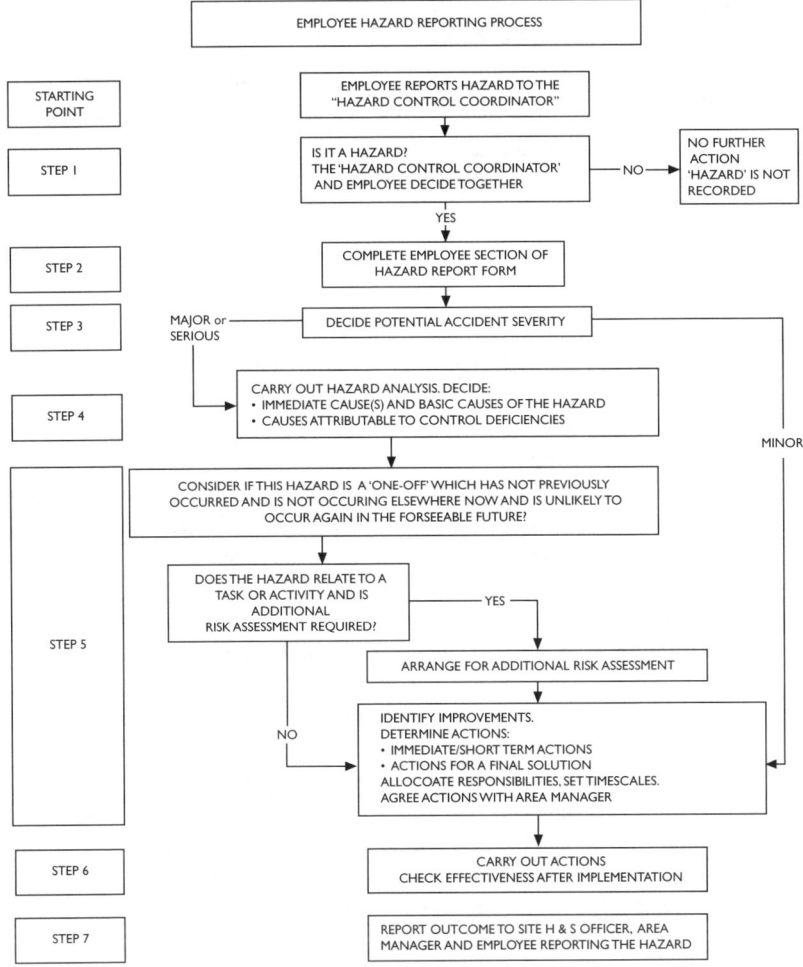

FIGURE 5.10

5.89 What also needs to be taken into account is the management challenge of ensuring that PPE is worn, as required, especially where the risk exposure is high.

5.90 Where it is not practicable to fully control the risk exposure other than by PPE, then the organisation must provide employees and others, eg visitors, with effective and suitable PPE to meet the needs of the job and the individual. Where an organisation provides PPE, it must do so free of charge to the employee. In the case of safety footwear and prescription glasses (either safety glasses or those provided for DSE use), many organisations will provide the item free of charge based on a standard issue. However, if the employee wishes to, they can purchase an item of their own choice (subject to the item meeting the company and legal requirements) and the company will make a contribution to the monetary value of the standard issue. A record of the issue of PPE will be kept.

5.91   Despite the fact that PPE should only be used as a 'last resort' control, it is still important to ensure that an effective and efficient system for PPE is used. As with all business-related processes, the process to manage PPE should be structured and consistently applied.

Process

5.92   The elements of an effective PPE process should be as follows:

- **Provision of PPE** – decide what is required and why. Create standard requirements for either general or specific applications. Decide which items of PPE are to be recorded when issued, re-issue frequencies, and early replacement policy.
- **Training for the use of PPE** – provide adequate instruction and training in the reasons for the issue and use of the PPE, with particular reference to the injury and ill-health consequences of not wearing the required PPE, plus potentially disciplinary sanctions (this last point is itself a last resort).
- **Maintenance of PPE** – ensure that the requirements for cleaning, inspection and storage are clear and are included in the above training.
- **Use of PPE** – the training should ensure that the wearer is able to fit the PPE so that it provides the protection required. Fit-testing for certain respiratory protection equipment (RPE) may be required.
- **Specific PPE** – any specific requirements should be clearly documented, including additional medical checks and linked to the Rules for a specific activity, eg a requirement to wear RPE during confined space working.
- **Purchase of PPE** – the general and specific requirements should include the actual specification of PPE to be purchased, eg type and strength of safety footwear, protection level for safety glasses, preferred suppliers for respirator cartridges, etc, so that the purchasing function will be able to source the required equipment and not primarily focus on other criteria, such as price when making a decision. Generally, all suitable and sufficient PPE for use within the EU will be 'CE' marked to demonstrate conformity.

Selection, standards and provision

5.93   Where PPE is required, the organisation should create a set of standards to cover:

- Employees, contractors and visitors.
- The wearing of PPE, by either general area or work activity or task.
- The methods for provision, cleaning, storage and replacement.

5.94 Line management and employees should be actively involved in the selection of the type and style of PPE to be worn. Typical selection criteria to be referenced are:

- Manufactured by a registered company.
- Complies with local or international standard, eg CE mark.
- Designed for type and degree of control required.
- Designed for activity intended and reasonably comfortable when worn under all situations where it is likely to be needed.
- Can be worn in conjunction with other items, ie compatibility.
- Durable, capable of being disinfected and cleaned (unless disposable).
- Ease of use and replacement of renewable parts.
- Acceptable to line management and wearers.

5.95 The PPE detailed in the organisation's standards must be readily available to those designated as wearers and appropriate storage and cleaning facilities should be provided to ensure the issued PPE remains in usable condition.

5.96 The process should be reviewed on a regular basis to ensure that it remains valid and that the requirement schedules are up to date.

Training

5.97 The standards and provision rules should be used as the base to provide adequate information and instruction to those who are required to wear PPE. The areas to be covered are:

- Need for and use of.
- Correct fitting.
- Cleaning, maintenance and replacement of parts.
- Reporting defects in the PPE.

5.98 Training can be given in a staged process, starting with general induction, departmental induction and work activity or task instruction.

Planned preventive maintenance (PPM)

5.99 Where replacement parts are included in an item of PPE, the PPM process should include a regular check and replacement of those parts, either when required by the manufacturer or local legislation. Replacement of parts should be recorded.

Compliance

5.100 Line managers are responsible for compliance to the standards within their area of responsibility and for highlighting sub-standard practices in all other areas.

5.101 Employees are responsible for ensuring that they conform to the standards for wearing, cleaning, storage and replacement of whole items and parts where required.

5.102 The organisation's hazard-control processes, especially planned general inspections, should be used as a periodic check on compliance.

Records

5.103 Records must be kept of the training provided to wearers and others.

Example general PPE schedule

5.104 Set out in the following table is an example of a 'site' PPE schedule that provides general guidance about PPE required. An actual schedule for the 'site' should be created following appropriate risk assessments.

5.105 The principle of using schedules is relevant and can be applied even to an organisation operating on a single site. The selection, purchase, issue, use and maintenance of PPE are not something that can be left to chance. It has to be very effectively and consistently controlled.

Planned preventive maintenance

5.106 It is vital to the management of OSH risks that an effective, efficient and proactively responsive process for the management of planned preventive maintenance (PPM) is in place. Additionally, the compliance to OSH policies and procedures by those who are directly responsible for managing the activities, plus those who actually undertake the PPM activities, will provide a valuable role model for all others in the organisation.

5.107 In some organisations those responsible for managing, and those responsible for undertaking, PPM activities, are very protective of their function, are often obstructive to those seeking solutions to problems, and often exhibit what we call a 'Tarzan' type approach to their activities. For example, an often-heard comment would go something like: 'Normal rules don't apply to us; we can access that position and make an effective repair without the need for the required equipment or safe working (for normal operations), which will only slow us down or get in the way anyway. Do you want the job done now or "next week", it's your choice, as we have plenty of other things to be getting on with?'. Management and OSH professionals cannot allow a two-tier system of compliance – one for normal day-to-day operations, and one for PPM activities, where the latter is less controlled than the former. Indeed, in many instances PPM activities may need a higher level of control, including safe system of work, permits to work, etc, than normal day-to-day operations.

Hazard/risk	Item of PPE	Frequency of issue	Training frequency/type	Record of issue?
Handling compressed gases, flying objects entering from front or side	Eye protection – glasses with side shields	As required	Line manager on first issue	No – unless prescription issue
Using liquid and solid hazardous materials, hot work or light radiation	Chemical goggles, welding masks, UV glasses as appropriate	As required and based on local/regional regulations	Ditto	Yes – if personally issued
Using corrosive liquids or solids, operating a grinder, etc	Face shield	Ditto	Ditto	Ditto
Falling objects or liquids and restricted height and customer requirements	Head protection	Ditto	Ditto	Ditto
Using cryogenic liquids or extreme temperature, sharp objects, corrosive materials or electrical charge	Loose-fitting leather gloves for cryogenic liquids and impermeable gauntlet gloves for corrosives	Ditto	Ditto	Ditto
Work equipment or other items falling on or trapping the foot	Protective footwear with appropriate strength toe cap and metatarsal protection when handling gas cylinders	Ditto	Ditto	Yes

continued

Hazard/risk	Item of PPE	Frequency of issue	Training frequency/type	Record of issue?
Potential exposure to noise above required level, eg 85dB	Ear plugs or ear defenders	As required – ear defender seals to be checked regularly for wear and reduced effectiveness	Line manager on first issue	No – unless prescription issue
Exposure to hazardous liquids, general cleaning agents, coolants, waste products, dirt, grease, etc.	Body protection, eg full overalls. Flame resistant when handling or working with pyrophoric gases, oxygen etc. For hazardous materials, a PVC apron (with sleeves) or coated chemical suit	As required – checked regularly for wear and tear and puncture holes	Line manager with local/region H&S specialist on first wearing	Ditto
Inhalation of toxic gases, vapours or by products	Respiratory protection equipment – half-face or full-face respirator, with appropriate cartridge, or full breathing apparatus or supplied air	As required	Specialist training prior to final wearing. Then at regular intervals based on local regulations.	Yes – if personally issued

5.108 If, however, the PPM function adopts a proactive and co-operative approach and is 'seen and heard' to be supporting the management of OSH and does not 'cut corners', then a different role model is created. Getting those responsible for managing and undertaking PPM activities 'on side' with OSH management systems will make a significant contribution to the general acceptance within the organisation that OSH management is part of normal day-to-day activities.

5.109 One other risk control measure that is particularly effective is to make the PPM function responsible or at least actively involved in the management of external contractors. In that way the internal OSH management system requirements will be consistently applied and managers will not be able to opt for an external 'low cost option' where OSH controls are minimal. We will return to this aspect below.

Policy

5.110 Organisations should ensure that an adequate preventive maintenance system (PPM) is in place to ensure that no work equipment or location/areas over which they have control present an OSH hazard/risk. A list of required OSH-related PPM should be created and maintained by the OSH professional, in conjunction with the internal function or external contractor(s) who undertake the PPM.

5.111 Details of required statutory inspections should also be incorporated in the PPM process.

Process

5.112 The organisation should develop a comprehensive process that creates and manages a schedule of PPM on all items of equipment, plant, tooling, etc, that require a visual inspection, regular lubrication, servicing, parts replacement or statutory examination. The process should record all planned and unplanned work to review the need to amend the schedule. The PPM process is crucial to the establishment and continued effectiveness of the OSH management system.

5.113 A systematic survey/analysis, using data from the following sources, should be used to identify and assess all items that potentially need PPM:

- All areas, processes, machinery, plant, other equipment, tooling, etc.
- Maintenance requirements defined by suppliers of items used for work activities.
- Maintenance and inspections required by legislation and/or risk assessment.

- Plant histories and PPM records.
- Items that exist especially for OSH, eg fire detection and protection, emergency equipment and health hazard control equipment.
- PPE that requires regular inspection, testing and maintenance.

5.114 The results of the survey/analysis should be used to create a PPM process, which includes:

- A description of the PPM process and responsibilities.
- A listing of *all* items requiring PPM.
- The method and frequency of PPM on each item.
- A system of work planning and record-keeping related to the frequency of PPM on each item.
- The recording of planned and unplanned work.
- A link to the 'Business process and operational changes' process.

Unplanned actions

5.115 Situations can arise where work is required (by internal maintenance personnel or an external contractor), on items that are included in the PPM process, but the work is outside of a scheduled PPM activity or on items that are not included in the PPM process. In these cases, the PPM records should detail the work undertaken and an assessment made of the cause(s) of the situation that led to the work being required. This information should be used to determine whether the PPM schedule needs amending, and if the work was required as a result of a situation that has OSH implications, an action should be entered on the organisation's actions and follow-up process (see above) to determine the implications and make appropriate changes to other parts of the organisation's processes, eg change of work instructions, change of purchasing specification.

Reviews and update

5.116 An annual review should be undertaken to ensure that the schedule is complete and/or the frequencies/PPM method need amendment in the light of experience and unplanned items repaired during the previous year or a change in circumstances.

Business process and operational changes

Introduction

5.117 Organisations should have in place a process for the formal review and approval of all projects that change the current methods of working or potentially introduce new hazards or risks. New materials and services, eg chemicals, equipment, plant, tooling or contractors, either required by these projects or via routine purchasing decisions, shall also be subject to a formal review and approval process to prevent an increase in hazards and risks.

Policy

5.118 Organisations should ensure that changes in business process and operations that affect workplaces are fully considered prior to implementation, with particular regard to OSH risks. The policy is to identify, discuss and agree (prior to implementation) changes in business process and operations that have OSH implications. In addition the process should cover the assessment of and approval to purchase items, eg materials, equipment, contractors that are required to implement/maintain the change.

Process

5.119 The organisation's process should ensure that all equipment, materials and processes are assessed in order to meet the organisational and legislative requirements for OSH and risk management, prior to approval and implementation of the change. The process shall cover all changes that affect the organisation's products/services, production/manufacturing processes and other supporting processes and facilities. Any process-related or design changes should also be included to ensure that hazards and risks are not increased.

5.120 The process applies to all employees involved with changes in business process and operations and the subsequent notification/communication of those changes to all concerned/affected by the change.

5.121 Line managers and those managing 'change projects' are responsible for ensuring that their activities and those under their control comply with this process.

5.122 The process should include:

- Definitions of items being changed or newly introduced.
- A description of the process steps and, if possible, flow diagrams.
- Standard checklists to guide the assessments.
- A link to the output from OSH and other risk assessments, health hazard surveys and systematic surveys.

5.123 Those with responsibilities for OSH, quality and the environment should approve the assessment before the change can be implemented.

Approved lists

5.124 When an organisation commences this process it will already be using many items of equipment, substances and services/contractors. Consequently, it is important to create a base from which to work by creating the concept of 'approved lists'. Items to be initially placed on the respective approved lists should be previously approved equipment, materials and services/contractors. Any item, including a service/contractor that is not currently on an approved list, should be regarded as 'new' and therefore has to be assessed for inclusion before purchasing can commence.

Purchasing

5.125   The organisation's purchasing process should include a check that a formal assessment of non-approved items has been completed and the purchase of the item is approved. A checklist should be used to ensure that all relevant checks have been completed. The checklist shall be counter-signed by those with responsibilities for OSH, quality and production. Only when the purchase has been fully authorised can the item be added to the appropriate 'approved list' and the purchase can commence.

Contractors

5.126   The controls for business process and operational changes should include:

- The creation of specifications of the work to be carried out by external contractors.
- The assessment and selection of contractors for the specific activity or task, including a check on the adequacy of the contractor's insurance cover for the work and methods to be used.
- An agreement on the detailed 'method statement' for the work to be performed.
- An agreement on the method of contractor/sub-contractor liaison and control during the work.

5.127   Only when these steps have been agreed between the contractor and the organisation should authorisation be given for the work to commence. The above list of items should form the basis of a written agreement between the contractor and the organisation, which should form part of the organisation's purchasing terms and conditions.

5.128   The manager with responsibility for the work shall ensure:

- That a formal induction on relevant aspects, including OSH, is provided to the contractor and their workers before they commence work.
- The contractor and their workers are fully supervised as per the agreement and that they comply with all Rules, especially permits-to-work.

- That the work is completed to the satisfaction of the organisation.
- That any clear-up after the work is completed is carried out to the satisfaction of the organisation.

5.129   The organisation's invoice approval process should take account of the above factors when payment is being authorised.

Example flowcharts for 'business changes'

5.130   Flowcharts have been produced which show:

- An overall business management system flowchart (Figure 5.11) shows the top-level organisational process for managing any business or operational change, and where the input of OSH commences.
- A health, safety and environmental flowchart (Figure 5.12) shows the next level process for the interface with the top-level process, and how OSH provides input to the change. It also details the links to the 'third level' processes, eg specific risk assessment and contractor assessment and selection.

Work equipment and complex machinery

Introduction

5.131   The process for work equipment is covered by specific regulations entitled the Provision and Use of Work Equipment Regulations 1998 (known as PUWER) which creates a process to assess suitability for purpose of an item of work equipment and selection criteria to minimise risk. Work equipment is any item that is provided for use at work, and can be anything from a hand tool, to display screen equipment, to a large power press or to a road vehicle for use by an employee. Whilst specific regulations relate to certain items, eg a road vehicle, PUWER has an all-embracing coverage when assessing the suitability of an item as work equipment. For example, a passenger car provided to an employee has to satisfy regulations relating to its design, etc, but will still need to satisfy work equipment regulations regarding its suitability for the purpose the organisation intends. If the employee is a service engineer and has to carry large spare parts or actual customer product, the vehicle should be able to carry and store the items without affecting the handling of the vehicle and the safety of the service engineer, should the items move following an accident or the sudden movement of the vehicle.

5.132   Additionally, the service engineer needs to be able to load and unload the vehicle without risk. The organisation may then conclude that an estate car, with self-levelling suspension and a rear-load guard between the load space and the passenger compartment is more suitable than a saloon car.

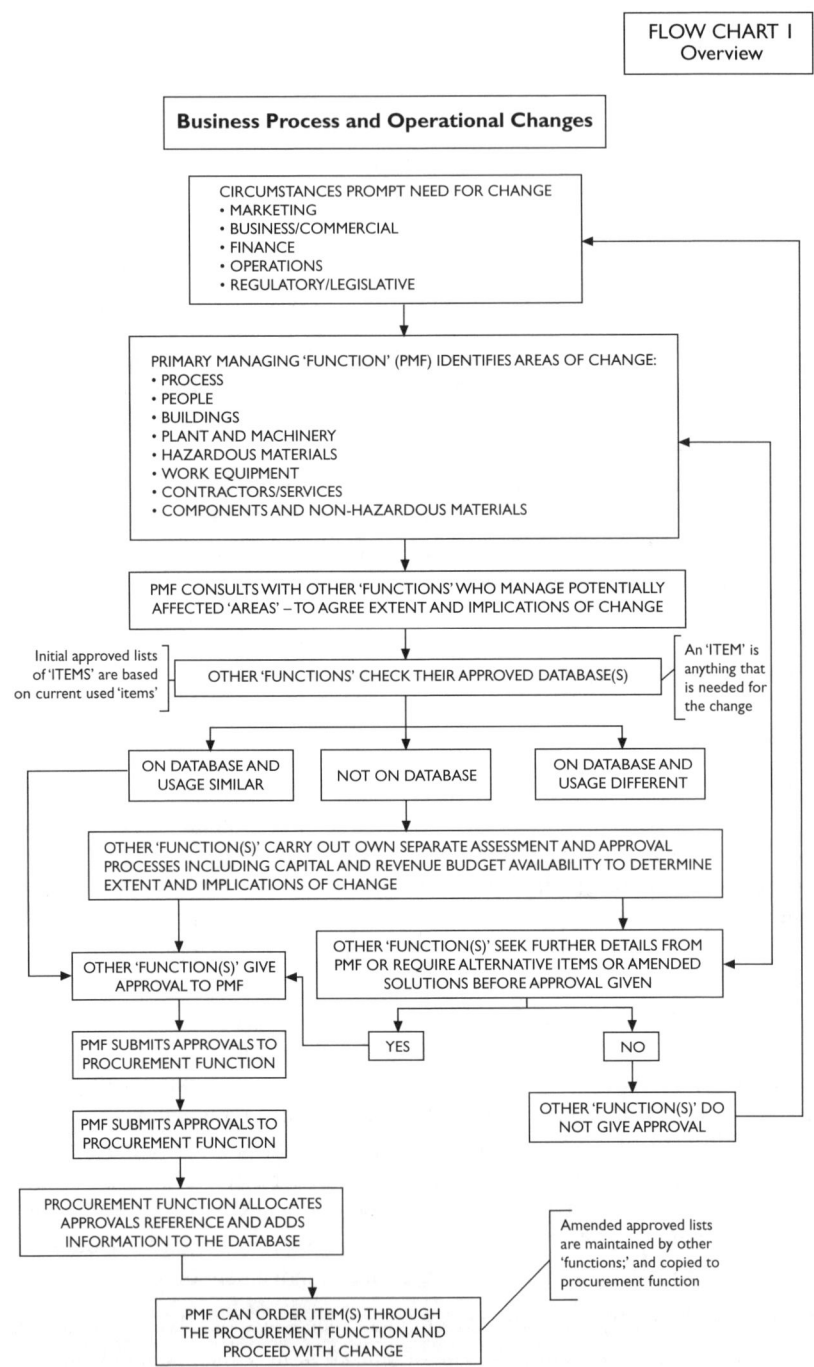

FIGURE 5.11

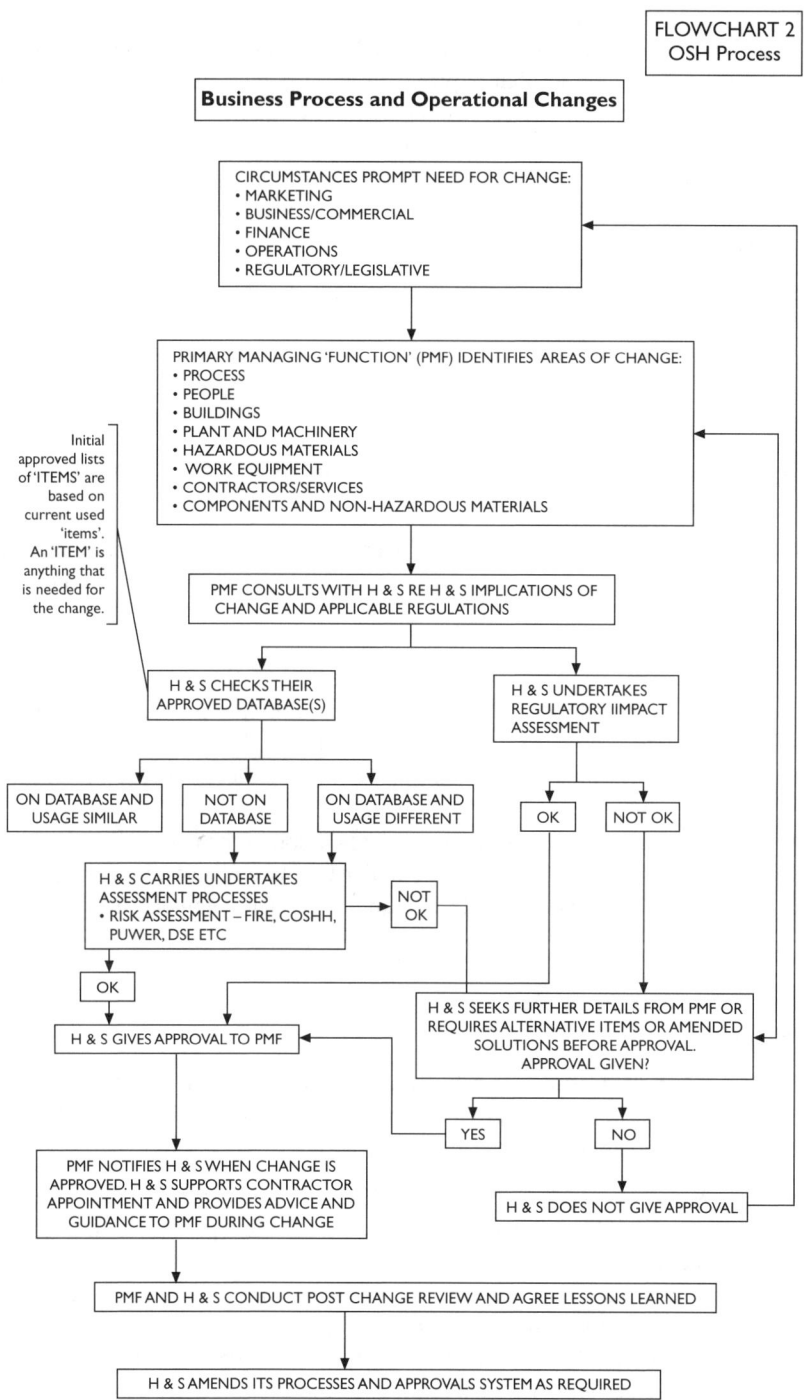

FIGURE 5.12

5.133 In addition the Lifting Operations and Lifting Equipment Regulations 1998 (known as LOLER), is linked to work-equipment requirements. LOLER covers any item that is provided to undertake lifting operations and lifting equipment. In many cases equipment covered by LOLER, eg a forklift truck is also work equipment, but each takes precedence where the requirements of either call for more risk control. For instance, a goods or passenger lift will be operated and/or used by employees, but is covered by LOLER for testing.

5.134 The process for work equipment and machinery has a strong link with 'Business process and operational changes'.

Policy

5.135 An organisation must establish a policy in respect of work equipment. Some suggestions include:

- Employees must not use or maintain an item of work equipment if they have not received adequate training to ensure they are competent in its use and/or maintenance.
- No manager will allow work equipment to be used for a task unless it has been assessed as suitable.
- Managers must ensure that appropriate training is provided for potential users of the work equipment.
- Managers must ensure that an adequate process is in place for the servicing and repair of work equipment.
- Managers must ensure that there is a comprehensive schedule of all work equipment that needs regular testing and servicing, and should ensure that the schedule is maintained. This includes lifting equipment, forklift trucks, etc.

Process

5.136 There is a need to establish a formal process for the identification, assessment, selection, implementation, training and maintenance of work equipment. Some general ideas are:

- Managers should ensure that the work equipment in their area of responsibility is authorised and is 'fit for purpose' and, where required, is listed under the PPM system for servicing and testing.
- Managers should ensure that all appropriate training has been provided to those using the work equipment.
- Managers should ensure that the work equipment is regularly inspected, and any faults promptly reported to those responsible for PPM.

- Managers should liaise with OSH and other functions when consideration is being given to the purchase of new work equipment.

Additional process considerations

5.137 Organisations can use the following list of good practice when developing their process for managing work equipment (the list is not exhaustive):

- Take all reasonable steps to ensure the OSH of all employees working with work equipment as well as to ensure the safety of others who may be affected by the work equipment.
- Liaise with suppliers to ensure that all new work equipment is designed and supplied to work in a safe manner and is suitable for the purpose intended by the organisation.
- Provide adequate information and training to all employees and managers.
- Carry out a risk assessment for all activities involving work equipment, and implement enhanced controls where required. Ensure that the risk assessment covers ergonomics, control systems, failures and misuses.
- Review the risk assessments on a regular basis and when process changes occur or production levels/methods change.
- Take effective measures to prevent contact with dangerous (moving, rotating, sharp, etc) parts of work equipment. The measures will either prevent access to the dangerous part, or stop the movement of the dangerous part before any part of a person can reach it.
- Ensure that the guarding is suitable for the method of working and provides adequate protection. Do *not* accept the assurances of the supplier that it satisfies PUWER; make sure that the guarding is discussed and agreed before the order is placed. If necessary ask to be shown examples of the guarding fitted to similar work equipment. In addition, all work equipment should comply with the EU Directive and so bear a 'CE' mark or have an EC Declaration of Conformity.
- Ensure that all PPM activities that involve risks to OSH can be carried out while the work equipment is shut down, or that appropriate protective measures are taken.
- Carry out risk assessments that identify specific hazards related to equipment failure. The organisation should take measures to prevent the hazard occurring or adequately control the risk.
- Restrict access to hot or very cold surfaces by engineering methods.
- Provide controls and control systems (particularly stop controls and braking systems) to assist with risk control, or to indicate that the risk has changed.
- Provide suitable means to isolate all work equipment from all its sources of energy. Isolation means establishing a break in the energy

supply in a secure manner, ie by ensuring that inadvertent reconnection is not possible. Safe systems of work, including permits to work, should also be considered.

- Ensure that work equipment or any part of work equipment is stabilised. Where stability is not inherent in the equipment design, operation or where external forces could compromise stability, eg adverse weather, vandalism, additional measures should be taken. Steps will be taken to ensure that mobile equipment is used within the limits of its stability at any given time.

- Ensure that suitable and sufficient lighting that takes account of the operations to be carried out, is provided at any place where a person uses work equipment.

- Ensure that work equipment is marked in a clearly visible manner with any marking appropriate for reasons of OSH.

- Ensure the equipment is suitable for the job and well maintained and that maintenance records are kept for all work equipment.

- PPE should be selected in accordance with the Personal Protective Equipment Regulations 2002 for use with machinery and work equipment.

- Appropriate mandatory safety signs shall be erected in accordance with the Health and Safety (Safety Signs and Signals) Regulations 1996.

- All fixed and portable electrical equipment will be tested on a regular basis and records kept. Equipment used externally to the organisation's premises must be marked as to its testing status. 110 v equipment must be utilised for on-site work.

- Inspect all work equipment after installation; after moving or modification work; periodically to assess wear and tear; after long periods of non-use and keep records of the inspection and, within the PPM system, keep records of the work undertaken.

- With regard to lifting equipment:
  - Ensure the lifting equipment is adequate in strength and stability under all conditions of use.
  - Ensure the lifting equipment is inspected in accordance with current statutory requirements.
  - Ensure the lifting equipment is marked with its safe working load.
  - Consider the effects of lifting objects or people.
  - Consider people who may fall, get trapped or crushed.
  - Workers should avoid being under working loads. If they need to be there, a safe system of work is needed if at all possible.
  - Lifting accessories must be suitable for the loads they carry.
  - Safe storage of lifting equipment is required.
  - Lifting equipment must have adequate planning and supervision.
  - Consider guidance (safe system of work) of non-guided loads, ie to prevent collision, tilting, moving, reversing, overturning.

- If a driver cannot see they must have a guide (banksman) who ensures safe movement of the load.
- Ensure a load can be attached and detached safely.
- Load must be safe in the event of a complete or partial power failure.
- Do not leave a load hanging unattended.
- Consider the effects of bad weather, high wind, etc.

People and vehicle risks

Introduction

5.138 Any person who drives as part of their work activities, either in a company supplied or personally supplied vehicle, is regarded as being 'at work' if the driving journey is work related. Indeed a vehicle supplied by an organisation is classed as a piece of work equipment (see above), but private vehicles are excluded from PUWER. If, however, the private vehicle is used for work-related driving, then the same requirements as under PUWER should be applied.

5.139 Organisations are required to undertake a risk assessment of the hazards and risks from work-related driving and incorporate risk-control measures within their OSH management system.

5.140 The most successful risk-management systems are based on an inclusive approach and not command and control, and these principles should be incorporated into a 'people and vehicle' risk management policy and system that relates to the organisation's business and risk profile.

5.141 In our experience, 'just doing driver training' does not produce a long-term sustainable improvement in the way that these risks are managed or the way drivers are exposed to these risks, or a reduction in the organisation's total costs. Communication and training can be important elements, but only as part of a long-term business risk management system.

Key success factors

5.142 The following factors are key to the effective implementation and sustainability of a people and vehicle risk management system that is part of your normal business and operational processes:

- Create a formal programme that involves:
  - Sizing the problem.
  - Setting a clear policy.
  - Clearly defining responsibilities.
  - Effective and sustainable communications.

- – Driver risk assessments.
- – Targeted training.
- – Regular monitoring and review.
- Integrate your fleet strategy with your people and vehicle risk-management system.
- Adopt a proactive risk management culture.
- Ensure committed and effective management attitude and leadership.
- Obtain proactive employee involvement, attitude and behaviour change.
- Establish strong teamwork, especially for field-based drivers.
- Create high visibility and awareness levels throughout the organisation.
- Use continuous improvement concepts.

Policy statement

5.143 A people and vehicle risks policy statement should be prepared and signed by the person who has signed the general statement of OSH policy for the organisation. It can either be a separate policy or included within the general policy. It should be used during induction processes, and included in employee handbooks and driver safety training. An example is given below:

'Our policy covers any employee undertaking a work-related journey which requires the use of a road vehicle, eg company vehicle holders; employees using a private vehicle for company business; pool car and hire car drivers. The organisation recognises that the activity of work-related driving needs to be managed within its overall management of OSH, and that vehicles supplied by the organisation, or other vehicles used by employees to undertake work-related driving activities, must be managed as a piece of work equipment.

Vehicles used for work-related driving are both a valuable item of an employee's work equipment and a vital element in an employee's ability to perform their job effectively. We jointly need to ensure that the vehicle used is in a roadworthy condition at all times, plus has all the required documentation. Regular maintenance and care in its use are important; as is the additional focus we place on driver safety and in particular good safe driving techniques.

Only authorised employees are allowed to drive company vehicles. Wherever possible, non-company car holders, or those with a regular work-related driving activity, should use either a pool car or a hire car if they need to undertake a work-related journey.

The use of a mobile telephone during a work-related journey is only permitted under the organisation's driving and mobile telephone use rules. Employees must report all accidents that take place in an organisation vehicle, pool car or hire car, even if the journey is not work related.

The organisation recognises its responsibility to promote driver safety in order to continually reduce accidents. Our long-term aim is to have no avoidable vehicle accidents. The organisation also recognises the need to reduce costs both in human and financial terms and to foster a professional organisational image by encouraging high standards of driving.'

5.144 Other policy considerations include:

- Management involvement – management should lead and control the development and implementation of the policy, which should be proactive in nature and based on an assessment of risk. It should concentrate on ensuring a positive attitude towards people and vehicle safety and ensuring the application of defensive driving techniques at all times, rather than concentrating on improving technical driving skills.
- Employee commitment – every employee who drives on a work-related activity is required to exercise and maintain their personal responsibility to maintain their vehicle in a safe condition and to apply defensive driving, techniques at all times, thereby protecting themselves and others potentially affected by their actions.

Risk assessment

5.145 The process of risk assessment should enable the organisation to understand where it is most at risk, and which drivers are most at risk, and guide the development of enhanced controls. We do not recommend that every individual driver is risk assessed using either desk-top and/or road-based risks assessments, especially where a large number of drivers are involved. Using a blanket approach, particularly for very low-risk drivers, will be wasteful of resources and de-motivate those who are very low risk.

5.146 In addition a targeted approach is easier to justify to those who are most at risk, because the evidence will be clear. Consequently, we recommend the use of the following graduated steps to identify the drivers most at risk, so that resources can be targeted where they will have most benefit in risk reduction:

1 Undertake an organisational review of the existing system for managing people and vehicle risks, and identify discrete driver groups based on risk profile – not necessarily based on vehicle type or job type.

2 Undertake a generic risk assessment for each discrete driver group, and use the results to determine where additional controls are required for whole groups.

3   Using the same factors, plus accident history, undertake a risk
        assessment of each individual driver within the most at-risk group(s)
        to identify the most at-risk drivers within the group(s).

4   For the most at-risk drivers use computer-based assessments, provide
        classroom-based   awareness   training,   and/or   road-based
        assessments/training to reduce their risk profile.

5   Use additional assessments where subsequent accidents take place,
        or on the recommendation of their manager.

Organisational assessment

INTRODUCTION

5.147   The purpose of the assessment is to evaluate an existing management
system, and identify current and historical risks and costs.

5.148   The assessment should look at the complete picture and all organisational factors that could affect the successful management of the risks. The
review may consider factors not normally associated with this area of risk.
Appropriate and available records are reviewed to identify the current situation.

5.149   Areas of review to include:

- Operations and vehicle usage.
- Fleet type, locations and employee work-related usage.
- Road risk management system and records.
- Fleet management policies and processes.
- Vehicle accident statistics.
- Vehicle accident management process.
- Direct and indirect costs.
- Type and level of risk.
- Current policy objectives and timescales.
- Management objectives and defined responsibilities.
- Current driver training.
- Safety culture.
- Management attitude and leadership.
- Employee attitude.
- Awareness levels.
- Interviews of a representative sample of work-related vehicle users.

REPORT

5.150   A detailed report is prepared, and a presentation made to senior management. Recommendations for improved control measures are included. The
report will identify the driver groups to be used for the next level of assess-

ment, and the risk factors to be included, plus the graduated scale to apply to each risk factor.

5.151   The factors to be used must reflect the organisation's overall risk profile and the type and level of its people and vehicle risks. Typical factors to be used are shown in the table below:

Risk factor	Description of risk factor
Work activity	Drivers undertaking routine activities at the end of a journey are less at risk than those who have to 'prepare' for the next visit whilst they are driving or whose journeys are changing at short notice
Journey frequency	Drivers who undertake occasional journeys are less at risk than those who are driving as a core activity
Journey distances	Drivers who drive short distances are less likely to be fatigued than those on longer journeys
Driving periods	Drivers are more likely to be tired working extended hours or on call
Traffic density	Drivers travelling on densely populated roads are more likely to be involved in a collision
Road profile	Drivers are less likely to have an accident on major trunk roads and motorways
Regular locations	Drivers in unknown locations are more likely to be distracted
Pedestrians	Hitting a pedestrian exposes the organisation to risk of civil action and reputation damage
Mobile telephone usage	Drivers who are frequent users are more at risk than those who limit the use of mobile phones
Driver age	Younger drivers are more at risk than older drivers
Driver length of service	Short-service drivers are more at risk than longer-service drivers

5.152   Using information from the organisational assessment, develop graduated risk categories under each risk factor. The graduated risk categories and a value for each can then be used to evaluate the risk level under each risk factor and the total for each driver group. The risk assessment matrix to be used should be consistent with the organisation's overall risk assessment process.

5.153 The outcome of the driver group risk assessments will indicate the groups most at risk, and start to identify the profile of those individuals within those groups that are most at risk.

INDIVIDUAL RISK ASSESSMENT

5.154 Undertake the same process to risk assess all the drivers in the most at-risk groups, plus factor in each driver's accident history. Use the outcome to identify those drivers who need additional assistance to reduce the likelihood of having an accident.

Management system

OBJECTIVES

5.155 The management system should ensure that managers and drivers:

- Understand and are able to implement their responsibilities.
- Are trained in order to improve and maintain their defensive driving techniques.
- Know how to take advantage of ongoing support.
- Are provided with vehicles appropriate to their job activities and, wherever possible, fitted with additional safety devices, eg self-levelling suspension for service engineers.

PROCESS

5.156 The main elements of a management process should be:

- Making managers aware of their responsibilities, the organisations' policy, need for action and the allocation of resources.
- The active involvement of line managers and employees in the development process.
- Employee co-operation with and application of driver safety policies and training.
- The assessment and selection of a driver safety training provider, who can provide:
  - Support for process development and implementation.
  - Training for managers and drivers, within the required geographical area.
  - Assistance to monitor the process and make recommendations to improve its effectiveness.
- The provision of theory and practical defensive driving training for managers and drivers.

- The inclusion of driver safety items in the recruitment and selection process for managers and employees who are to drive on work-related activities for the first time, to assess their suitability to drive on work-related activities. This should include requirements for minimum driving experience, limits related to driving offences and specific licence category requirements.
- The inclusion of driver safety in new manager and employee induction processes.
- The provision of vehicle and personal safety items to reduce hazards and risks.
- Line manager assessments of drivers, including, where allowed, checks on driving licences.
- Communications on driver safety to assist the maintenance of the correct attitude and behaviour.
- The inclusion of the subject in all meetings between a line manager and a group of drivers.

Vehicle assessment and selection

5.157 A process should be used to assess and select appropriate vehicles for each type of user or job. Line managers and drivers should be involved in vehicle assessment and selection. Factors to be taken into account include:

- Type and length of usage, eg short journeys in town or long urban journeys.
- Overall visibility, particularly for low-speed manoeuvres and parking.
- Load factor of items carried, eg service engineer spare parts.
- Availability of restraint systems for passengers and loads.
- Standard safety features, eg side impact bars and ABS.
- Availability of additional safety items, eg self-levelling suspension.

Driving assessment and training

5.158 It is vital that any driver assessment and training interventions are undertaken by a training provider that works in partnership with the organisation. As with any training intervention it is necessary to clearly identify what are the objectives and select methods that are most effective. Simply accepting that typical theory sessions and practical road sessions will solve any risk problem misses the fact that interventions must be tailored for the organisation.

5.159 The organisation must select a driver training provider based on its needs – not what the provider can provide or what they tell the organisation what is needed. This is especially important if the organisation needs to use more than one provider because of different vehicle types or geographical considerations where drivers operate in more than one country.

RECORDS AND MONITORING

5.160   Full records of all main process activities and training shall be kept. Line management in conjunction with the approved driver-training provider should regularly review process activities to ensure they are maintaining their effectiveness to reduce hazards and risks for drivers.

Accident management

Introduction

5.161   The organisation should have a formal documented process for the notification, investigation and reporting of all accidents/incidents. Those with responsibility under the process should receive adequate training. The organisation should also create a process for the detailed analysis of accidents/incidents and feed that information into a hazard control process detailed above. The process should cover all work-related accidents, including vehicle accidents, whether in vehicles supplied by the organisation or not. There is a legal requirement to report certain types of accidents (the Reporting of Injuries, Diseases and Dangerous Occurrences Regulations 1995) but there is no legal requirement to investigate accidents.

5.162   However, we strongly recommend that all accidents/incidents are investigated to a degree based on the seriousness or potential consequences of the accident/incident. Without the data from investigations, the organisation is not in a position to understand the obvious causes and the underlying causes. The purpose of investigation is not to allocate blame, but to identify the actions needed to prevent a reoccurrence. In addition, we recommend that the results of the investigation be recorded, again to a level of detail based on the seriousness, etc, of the accident/incident.

Policy

5.163   The policy should:

- Provide an effective system of accident management to protect the health and safety of employees, the environment and those affected by the activities of the organisation.
- Implement and maintain an effective system of accident management for reporting on, responding to, communicating about, investigating, following up on and analysing all work-related accidents, diseases and dangerous occurrences that result in, or have the potential for, loss. The level of reporting and investigation on each accident/occurrence will depend on the actual or potential level of loss, and regulatory requirements.

- Conform to regulatory and internal requirements regarding the reporting of incidents and accidents on organisation premises.
- Analyse the data from incident/accident reports to enable improvements in control to be identified and implemented.
- Use the organisation's process for the reporting of, and adequately investigating, all work-related incidents and accidents to ascertain the causes, lessons learned and remedial actions. The particular items to be reported and investigated are:
  - Injuries and fatalities to employees, customers or third parties.
  - Motor vehicle accidents.
  - Actual or potential damage to plant, equipment and buildings, including fire and other emergencies.
  - Spills or leaks that cause actual or potential harm to the environment.
  - Breach of a legal requirement likely to lead to legal action.
  - Adverse media coverage.

Process

5.164 The process should:

- Allocate general and specific responsibilities.
- Detail management participation – at the scene and thereafter.
- Define the accidents/incidents to be notified, investigated and reported.
- Include the internal reporting forms and distribution.
- Describe the investigation and review procedures, including the use of the advanced techniques, eg causal tree method.
- Describe the handling of post-accident situations, eg treatment and rehabilitation of injured employees.
- Link to the organisation's hazard control process and the reporting on the progress of actions arising from accidents/incidents.
- Classify accidents/incidents for internal purposes.
- Allocate responsibilities for external reporting to regulatory and other bodies.
- Link to the organisation's core training needs process to identify those with a training requirement under this process and its provision.
- Ensure that there are adequate records of each accident/incident that provide sufficient detail to enable retrospective analysis and are in a readily accessible format. Records should be kept for at least three years, or longer if prescribed by regulation and/or insurance policy conditions.
- Allocate responsibilities for the comprehensive analysis of all accidents/incidents, which includes the following:

- The identification from all accident/incident reports of the:
  - Related situational factors.
  - Underlying causes – immediate and basic causes and lack of programme control.
- The analysis of the extracted data, conclusions and recommendations.
- The calculation and circulation, every three months, of the key frequency rates or other analysis, based on the type of accidents/incidents that have taken place historically or in the period.
- The reporting, at least every three months, of the results of the analysis to line managers and the organisation's OSH committee for discussion on corrective actions.
- A link to the organisation's hazard control process.

Monitoring

Introduction

5.165 In **Chapter 4** we described in detail the reasons behind monitoring and how it fits within the P O P I M A R factors. A business process for monitoring the performance of the OSH management system needs to be firmly linked with the organisation's normal management system and a key element of the organisation's and individual manager's performance assessment. Account should be taken of the organisation's structure and the allocation of responsibilities for OSH management.

Policy

5.166 The purpose of the monitoring process is to provide regular information and undertake regular reviews and audits to determine the application of the OSH management system, the organisation's performance on OSH management, and progress towards its organisational objective for OSH management. The monitoring process should also produce recommendations for corrective actions. The main focus should be the 'work to be done', but details of 'consequence' measures should also be included, with the information biased towards 'lessons learned' rather than numbers of accidents/incidents.

Process

5.167 The actual monitoring elements must be co-ordinated, and similar in concept to the hazard control process, graduated to produce sufficient detail to inform the levels above or below the element being used. Typical monitoring elements are:

- Monthly – line managers to report on OSH performance in their business performance reports.
- Monthly – the organisation's OSH professional to provide an overview report to senior management.
- Bi-monthly – the organisation's OSH professional to report to the organisation's most senior OSH committee on system status and ongoing activities.
- Quarterly – the organisation's OSH professional to undertake a vertical audit of one or more OSH management system elements in a part(s) of the organisation. The audit schedule should be published in advance. Details of the outcome should be provided to the part(s) of the organisation audited, with an overview included in the OSH professional's report to the OSH committee.
- Site visits – if the OSH professional is responsible for more than one site, then at least once a year they should undertake a status review of each site in conjunction with the most senior management at the site. The status review will be limited to a small extract from each element of the whole of the main audit programme, and the report should indicate whether progress has been made; whether progress has been static or whether progress has been reversed, ie has the site moved forward, stood still or gone backwards?
- Annually – an external body should undertake a formal audit of the OSH management system. The content of the audit, its coverage and the type, ie horizontal or vertical or a combination of both, should be based on the results of other monitoring activities during the year, the results of the hazard control process, and the accident/incident performance.

CONCLUSION

5.168 The implementation of the above business processes are designed to ensure that OSH risk management requirements are integrated with normal organisational processes. If OSH processes are established as part of the way that the organisation runs its activities, then OSH will be managed as a normal part of management and employee activities. Management and employees must be heavily involved in the design, implementation and ongoing monitoring of these processes. The organisation's OSH professional should not be required to be the management system's policeman, but should focus on advising management and employees, in order to minimise the need for constant 'firefighting', thereby resulting in proactive resource allocation to ensure continual improvement.

CHAPTER 6

Organisational and human factors

INTRODUCTION

6.1 This chapter discusses how organisational psychology can provide a valuable perspective in the management of organisational risks. This approach to risk management has been evolving into a unified field over the past 15 years and is now becoming a key perspective in understanding the causes and motivations associated with risk related behaviour in the workplace. 'Organisational' and 'human' factors are discussed as part of an integrated approach to risk management. Organisational factors are the structures and processes that influence the culture of an organisation and include responsibility frameworks, communication frameworks and job/role design. Human factors describe fundamental human characteristics that need to be managed as part of the risk management process. This section includes an analysis of the perception of risk, the nature of human error and the motivations that drive individuals to take risks in the workplace.

6.2 It is argued that OSH professionals operating at a strategic level are well placed to play an influential role in enabling organisations to manage the risks associated with organisational and human factors because, essentially, risk management is about implementing the procedures and practices that lead to changes in behaviour, attitudes and values. The cross-functional nature of the OSH function provides an opportunity for OSH professionals to become an integral and vital part of the risk management process.

6.3 The management of risk related behaviour via an examination of organisational and human factors provides an invaluable focus for organisational development because developing systems to avoid human error and manage risks necessarily results in the development of quality standards. A systematic focus on limiting human error and controlling risk acts as a mechanism for organisational development, as the potential for human error is largely synonymous with inefficient organisational systems.

6.4 The main purpose of this chapter is to explain:

- How organisational and human factors result in error and risk related behaviour.
- How and why OHS professionals should manage organisational and human factors as part of an integrated risk management system.
- How the management of risk leads to organisational development.

INTRODUCTION TO ORGANISATIONAL FACTORS

6.5 Organisational factors are the systems, procedures and practices that ultimately determine the culture of an organisation. They influence the values, beliefs and expectations that members of the organisation come to share. Implementing an effective risk management system will require an understanding of how organisational factors affect risk related behaviour. Designing organisational factors that minimise the risks of loss and accidents, maximise intelligent risk taking to develop opportunities and better control the uncertainties associated with being in business will, over time, result in the creation of a positive risk culture.

6.6 Where organisational factors are not aligned to minimise risk, latent failures may occur. Latent failures are an accumulation of uncontrolled risk such as a lack of clear responsibility for job/roles, poor communication processes and a lack of training that come together unexpectedly to create failure and loss. Catastrophic organisational failures are often the result of poorly managed organisational factors.

6.7 The organisational factors referred to in this book are listed in **Chapter 1**, and used for a risk identification process in **Chapter 7**.

6.8 This chapter concentrates on organisational factors that are believe to be more significant in creating the structural framework of an organisation and therefore provide a focus for organisational development. Implementing an integrated risk management system is essentially an organisational-development process. The organisational factors discussed are:

- Senior management commitment to the risk management process.
- Organisational cultural and risk management.
- Responsibility frameworks.
- Occupational stress as a risk factor.
- Job/role design.
- Communication frameworks
- Learning and development

SENIOR MANAGEMENT COMMITMENT TO THE RISK MANAGEMENT PROCESS

6.9 No organisation can eliminate all risks, nor is it sensible to do so. Almost all human activity involves risk; indeed, the advancement and the improvement of society necessarily involves the evolution of new approaches that may result in exposure to new risks. Surviving and thriving in a business environment is much the same. Organisations are required to make continual progress, for example, by finding new ways of using resources, becoming more efficient and more competitive, developing new product niches, utilising new technologies, etc. The very business of being in business involves risk. The key is to ensure that the management of risk is a balance, between the maximisation of opportunities and the minimising of risk. What needs to be avoided is to engage in a risk averse approach that stifles the creativity of the individual within the business and therefore the organisation itself. Intelligent risk taking is a prerequisite for the development of a successful organisation.

6.10 All organisational change involves risk, and failure to change in a rapidly changing business environment also entails risk. Risk management is essentially good management and therefore requires a commitment from senior management to identify and manage risks, be they current business process and operational risks, innovation risks, or risks associated with deep-seated organisational and human factors that could result in organisational failure and loss. If responsibility for risk management is taken at senior level, this will help to ensure that those responsible for implementing strategy will have the necessary support, authority and resources.

6.11 A commitment to establishing a risk management process requires, in the first instance, that the business and human case for implementation are clearly understood and communicated to all levels within the organisation and to all stakeholders including external agencies. The benefits derived from implementing a risk management strategy include the traditional benefits of avoiding hazards, minimising the likelihood of failure and the general prevention of loss. However, as is discussed throughout this book, effective risk management also has other important positive outcomes. These include enhancing performance, improving communication systems, providing opportunities for organisational development and allowing the organisation to expose itself to greater business opportunities because existing organisational risks are better controlled and the uncertainties associated with new business development are better understood. Changing the culture of the organisation so that the management of risk becomes a core capability will, over time, lead to the following specific benefits:

- An enhanced organisational culture that develops the capability for balancing the maximisation of opportunities and the minimisation of risks.
- Increased likelihood that organisational objectives will be achieved as the discipline of risk management ensures that potential risks are identified, opportunities maximised and losses minimised.
- Enhanced shareholder value as the rewards derived from engaging in managed risks are realised.

- Better communication as risk management will require an organisation-wide framework for communication and a common language to discuss risk.
- The promotion of improvement via a risk management cycle that continuously reviews the organisation's responses to system failures and human error.
- Fewer negative unforeseen circumstances as the organisation becomes better at forecasting risks and potential outcomes.

6.12 The foundation on which successful risk management is based therefore comprises of a senior-level responsibility for risk management because there is a clear understanding of its benefits. Once this is in place, risk management strategies can be aligned to organisational objectives and integrated with the strategic planning process.

6.13 Some commentators have argued that the lack of senior management commitment to (safety and) risk management is 'more perceived than real' as senior managers are intimately involved in business processes, operational processes and innovation risk management, albeit often in an unstructured manner. Senior managers are also acutely aware that poor risk management can potentially impact on corporate profits and reputation, particularly if the stakeholders' view of the organisation were to become tarnished by what would be ultimately perceived as poor management. These commentators suggests that barriers to risk management are largely created by the resistance of middle managers to focus on risk management issues, perhaps with some justification as the role for which they are employed is primarily concerned with meeting 'production' targets. Although they may hear the rhetoric from senior management of the need to manage risks, they are confronted by other, often stronger and more urgent messages such as the need for cost-cutting, driving up 'productivity' and meeting business and commercial objectives. From the perspective of this book however, it is argued that effective senior management necessarily involves the management of middle managers to manage risk. The key to developing a positive risk management culture is therefore to implement an effective risk management strategy that has senior management support, and the active involvement and accountability of all managers.

ORGANISATIONAL CULTURAL AND RISK MANAGEMENT

Background to organisational culture

6.14 The concept of organisational culture is widely acknowledged to be very useful in helping to understand an organisations identity and performance aspirations. Schein (1990) defined organisational culture as 'the values, beliefs and expectations that members in an organisation come to share'.

6.15 Organisational culture is influenced by, and evolves as a result of, a number of interrelated variables such as the organisation's procedures, goals, management style, structure and human capital management. Despite the negative connotations of the term 'human capital management', ie the notion that individuals are resources that can be bought and sold, the term seeks to emphasise that an organisation's people are a 'capital' resource, equally as important as financial capital. They are vital to the success of an organisation and should attract investment rather than being seen as a potential cost. Organisational culture is also related to broader variables such as the organisation's function, size, history and, ultimately, national cultural variables. Organisational culture provides employees with information and direction as to how they are expected to respond to organisational initiatives such as organisational change, innovation or developing risk awareness.

6.16 Organisational culture may be very homogenous, fragmented or there may be very little consensus. Organisations without strong cultural norms may develop subgroup cultures built around successful 'local' leaders or as a result of geographical or functional separation. The aim of most organisations is to develop a unified culture that is consistent with the strategic goals of the organisation. Assisting organisations to develop a positive risk aware culture is one of the aims of this book. The extent to which there is a consensus among employees in how an organisation is perceived, particularly with respect to subcultures, relates strongly to measures of organisational climate. Organisational climate, as distinct from organisational culture, attempts to identify and measure an employee's subjective perceptions such as morale, motivations and attitudes rather than the values, beliefs and expectations that organisations seek to inculcate.

6.17 The value of attempting to measure organisational culture and climate has been increasingly acknowledged over the past 15 years because of its usefulness in mobilising employee effort through the creation of 'appropriate systems of shared meaning' (Morgan, 1997). The successful development of organisational culture has wide implications for the effective implementation of planned interventions such as encouraging new entrepreneurial styles, coping with mergers and acquisitions or embedding a risk management system. It attempts to create or impose a sense of corporate identity, by seeking to generate in employees a sense of commitment to goals that are larger than individual self-interests or the interests of functional groups.

6.18 Morgan (1997) highlights the fact that 'since organisations ultimately reside in the heads of the people involved, effective organisational change always implies cultural change'. An organisations is at least as complex as the employees that make it up, as people will have their own agendas and perceptions based on their own personal histories and motivations and these may be different to those promoted by the organisation. Therefore simply changing policies, procedures and technologies will not immediately result in a change in culture, because culture lies in the perceptions, attitudes and values of individuals.

6.19　Organisational culture evolves over time and is nurtured by appropriate learning and development programmes, the learning by good example and the use of rewards and sanctions that seek to modify attitudes and behaviour. It is sustained by a corporate vision that seeks to successfully manage and achieve organisational goals.

A positive risk culture

6.20　The term 'positive risk culture' has its roots in the concept of organisational culture. Essentially it describes an organisational culture that recognises consistent and effective risk management practices throughout the organisation.

6.21　There are numerous definitions that focus to a greater or lesser extent on organisational or human factors, but for the purpose of this discussion a positive risk culture is considered to be the result of an interaction between these two variables. This is because organisational factors are determined by human factors, i.e. the perception, attitudes and capabilities of those who design and implement the risk management system, and by the behaviour and thinking of employees that are influenced by organisational policies, practices and procedures. Human factors such as the perception of risk and motivational aspects of risk related behaviour are discussed in detail later in this chapter.

6.22　In order to be effective, a risk aware culture requires 'buy-in' by all stakeholders. Effective risk management results when there is a 'collective acceptance' of risk management practices and that this exists at all levels – individual, group, division and organisational levels, (Schein,1990).

6.23　Traditionally, attempts to improve risk management in the workplace have focused primarily on safety issues. They concentrated on the development of machinery to safeguard against injury and procedures to limit human error and accidents. However, a series of major accidents such as the meltdown of the nuclear reactor at Chernobyl, the fire at King's Cross and the explosion on Piper Alpha have highlighted the role that organisational culture contributed to each incident. Following the Piper Alpha inquiry, Lord Cullen stated that, 'it is essential to create a corporate atmosphere or culture in which safety is understood to be and is accepted as, the number one priority' (Cullen, 1990). More recently Lord Cullen's inquiry into the Ladbroke Grove rail accident in 1999 (Cullen, 2001) pointed to evidence that suggests that a large proportion of accidents, incidents and near misses occur as a result of a culture that did not seek to effectively manage risks.

6.24　As knowledge of the factors that lead to organisational failure and loss has developed, so it has become increasingly clear that, from an organisational perspective, there is little value in blaming such failures on chance environmental factors, technological failures or the behaviour of individuals who are responsible for triggering an incident. In the final analysis, responsibility for such incidences rests with those who are charged with running the organisation and the culture they develop to manage its risks.

Embedding a positive risk management culture

6.25 The first stage in creating a positive risk management culture is to identify appropriate strategies, policies, procedures and standards to manage risks. These are discussed throughout this book. Once these are in place, the embedding process results from allocating roles to these structures and assigning responsibilities. Roles and responsibilities can then be monitored, supported and enforced. In addition to the above, clear delegation of responsibility at all levels for key organisational objectives will create, over time, a culture where the management of risk is seen as an intrinsic part of how the organisation operates.

6.26 As people in the organisation change, it is necessary to ensure a continuous understanding of the roles and responsibilities related to the management of risk. The risk environment is also continuously changing. Therefore, as organisations adapt to a changing business environment or engage in new initiatives, priorities with regard to risk will also change. Assumptions about risk therefore require regular review and consideration. A common practice for organisations that manage risks effectively is to hold regular reviews of the risks associated with each of the key organisational functions. In this regard, the Combined Code on Corporate Governance, published in July 2003 provides a clear description of how organisations should monitor their internal control systems, including the management of risk.

6.27 The OSH function can play a pivotal role in monitoring risk related behaviour, providing suitable learning and development, and ensuring that performance management, compensation and rewards are linked to risk accountability. The involvement by OSH becomes particularly valuable if they have been involved from the outset in designing risk management policies, strategies, standards and practices. These aspects are discussed in greater detail in other chapters.

RESPONSIBILITY FRAMEWORKS

6.28 A risk management process requires an iterative cycle of identifying risks, assessing their impact and prioritising actions to control and reduce risk. Implementing this cycle requires that responsibility for each phase of the cycle at different levels of decision making is well defined. Because risk management is an integrated process across the whole organisation, it is important that accountability for specific areas of risk management is clear and that those responsible actually have the authority and capability to fulfil their responsibilities. It is also important that these roles and responsibilities are communicated and understood by others in the risk management process and the wider organisation.

6.29   Middle managers are a key group because, although they may have the authority to decide work practices, they may not have the responsibility for risk reduction or may not consider themselves accountable for negative outcomes. This can lead to a conflict of interest. As has been discussed above (**6.13**), productivity may be seen as incompatible with risk management, particularly when risks are perceived as insignificant. It is important, therefore, that the delegation of responsibility is linked to accountability. Shared or unclear lines of responsibility and accountability between individuals or functions will impact negatively on the risk management process.

6.30   In order to increase responsibility for risk management, especially at the operational level, middle mangers and employees should be involved in the identification of risk and risk assessment processes. This will lead to greater ownership of risk and help inculcate a positive risk aware culture.

6.31   The integrated nature of risk management will benefit from a group specifically established to facilitate and co-ordinate risk management across the organisation. This group can promote an understanding and assessment of risk, be responsible for reviewing systems and procedures and advise on the introduction of enhanced controls. Each organisation needs to decide on the most appropriate composition of such a group, taking account of their organisational structure, existing responsibilities, etc, but the following functions, as a minimum, should be considered for representation at the highest level: finance, risk/insurance, operations, OSH, a non-executive director and a number of operational-level managers.

6.32   Traditionally, the functions mainly involved in risk management are finance and insurance. They have tended to manage the risks associated with their function independently of the rest of the organisation. The implications of this type of 'silo management' are discussed in greater detail in **Chapter 7**. The argument presented in this book, however, is that OSH professionals operating at a strategic level are well placed to play an influential role in enabling organisations to manage organisation-wide risk because, essentially, risk management succeeds or fails depending on altering employees' perceptions, attitudes and behaviour with regards to risk. This involves learning and development programmes, performance management, reward and sanction systems and developing work practices and procedures that limit human error, increase job satisfaction and reduce stress. Because the above mechanisms are cross-functional and interrelated as is OHS, OHS professionals can take a lead in partnership with line managers to directly influence risk management practices. OSH cannot be marginalised without undermining the risk management process. In the UK, the Turnbull Report (1999) (a significant contributor to the thinking on the Combined Code on Corporate Governance (2003) mentioned above, (**para 6.26**) recommended that risk management should be imbedded in the operations of an organisation and form part of its culture.

OCCUPATIONAL STRESS AS A RISK FACTOR

6.33 Occupational stress has many negative consequences for organisations, and therefore it is a strategic risk that can be identified, quantified and managed. Numerous studies have shown that workplace stress adversely affects work performance, morale and commitment to the organisation. Workplace stress may cause absenteeism through physical or psychological ill-health, or an increase in turnover as employees leave the organisation for less stressful working conditions. This has significant implications for managing and developing an organisation's human capital, especially with regards to resourcing plans. In addition, workplace stress is implicated in reduced productivity and an increase in the likelihood of risk related behaviour that results in negative consequences. Negative consequences include increased costs resulting from poor quality work, increased errors and accidents and increased customer dissatisfaction. Where organisations have suffered catastrophic failures, failure to manage workplace stress is often implicated as a causal factor.

6.34 The existence of workplace stressors are costly to organisations and to society in general. Figures published by the UK Health and Safety Executive (HSE) (2003) estimated that work related stress costs employers about £380 million *per annum* (in 1995/1996 prices) and to society about £3.8 billion. Since these calculations were done, the number of days lost due to stress has been estimated to have more than doubled (Jones et al, 2003). The 'hidden' costs of stress to an organisation can be substantial and insidious because excessive workplace stress can create a dysfunctional organisation where the culture is one of resisting change and being unwilling to co-operate. In extreme circumstances a subculture may develop that actively seeks to undermine the organisational goals and feels justified in doing so.

6.35 The work stress literature is replete with advice and guidance on how to identify and quantify work-related stress and how to manage stress on a personal level, but until recently, there has not been much guidance on how to actually tackle the organisational factors that cause stress. The HSE has published for consultation a draft 'management standards' document on occupational stress. When finalised, it will assist organisations to tackle the challenges of reducing stress in the workplace. The HSE has recognised that many health and safety professionals are acutely aware of stress as a heath and safety issue but are not in a position to do much about it because tackling

the organisational factors associated with stress often involves major organisational reconstruction in work practices and would also involve challenging
deep-seated attitudes and management styles. In most cases those involved in
occupational heath and safety will not have the skill to undertake such large-
scale changes or the authority to do so. The publication of the 'Management
Standards' will highlight the role of 'general' management in tackling work
related stress. It is hoped that the standards will provide an opportunity for
OSH professionals to raise the profile of workplace stress with senior managers and to initiate debate on what needs to be done to address it.

6.36 Stress is caused by the interrelation of a number of human and organisational factors. OSH professionals can therefore do much to advise on the
modification of organisational factors to improve stress levels while also
improving productivity and job satisfaction. Consequentially, this presents
OSH professionals with a major opportunity to become actively involved in
managing the risks in this area. As work demands and business competition
increases, so organisations will need to seriously consider the moral, legal,
financial and business benefits of managing stress as a risk issue. OSH professionals can take a lead in the identification, evaluation and resolution of
work related stressors as part of their involvement in risk management.

Stress and stressors

6.37 Stress can be defined as a physiological, psychological and emotional
response to threatening situations or events. It results when there is a mis-
match between the perceived demands of a situation and the perceived abilities of the individual to cope with these demands. The implications are that
different people can interpret the same stressors in a different way, because
they have different perceptions of the demands and their abilities. Stress is
therefore a human factor but the causes of stress in the workplace, ie occupational stressors, are organisational factors. The focus of this section is on
stress as a stressor and is therefore included as an organisational factor.
Organisations can do much to redesign job/roles so that demands are achievable and employees feel they have the required abilities to meet the demands.
Stress is caused by the interrelation of human and organisational factors.
OSH professionals are therefore well placed to support the modification of
human factors such as the perception of risk via learning and development
programmes and organisational factors such as job/role design. This presents
OSH professionals with a major opportunity to become actively involved in
managing organisational risks.

Examples of improvement strategies that require OSH involvement include:

- Clarification of roles and responsibilities to remove stressors.
- Increasing individual job control and decision latitude.
- Building and supporting effective teams.

- Creating a supportive working environment.
- Ensuring focused learning and development to meet organisational needs and individual requirements.

Work overload

6.38 A common source of stress at work is being required to carry out more tasks than can reasonably be achieved in the time available or having to work at excessive speeds on a continuous basis. Work overload can also be qualitative; that is, work that requires excessive concentration or when employees do not have the necessary competencies to carry out the task without expending excessive effort. Stress arising from work overload is likely to be exacerbated by other stressors in the workplace.

6.39 A practice that is becoming increasingly common as a result of work overload is to take work home to complete. The long-term implications of this coping strategy are, of course, that occupational stress will leach into an individual's home life, thus allowing less time to recover from workplace stress with all the consequences that brings.

6.40 The drive to increase productivity, for example, by reducing head-count, has increasingly led organisations to make greater demands on their employees and this has resulted in workloads growing over time. However, increasing workload beyond a certain point does not automatically lead to an increase in productivity. An organisational culture that requires employees to exert excessive effort in carrying out their work is probably one that is poorly managed. Productivity is much more likely to be increased by well-designed job/roles and procedures, ensuring that employees have the appropriate competencies via effective learning and development programmes and ensuring that recruitment results in new employees with the required knowledge, skills, aptitudes and attitudes.

6.41 A risk management approach to reducing stress caused by work overload would therefore actively seek to identify workload stress areas as part of the performance management process or via a stress-rating questionnaire. Managers and employees can work together to make use of this type of organisational feedback to improve work flows, learning and development programmes and recruitment functions.

Role ambiguity and role conflict

6.42 A major source of workplace stress for individuals results from a lack of clarity about their responsibilities. Ambiguity is experienced as a lack of control in being able to set work priorities or being controlled by the inconsistent expectations and demands of others.

6.43 A related source of workplace stress is role conflict. Role conflict occurs when employees are asked to carry out roles that they perceive to be incompatible or when their roles are in conflict with their personal values and beliefs. For example, a manager may be required to canvas employee opinion on a contentious work issue while knowing that the exercise is merely playing lip service and employee opinions will be ignored.

6.44 The negative health effects of stress caused by role ambiguity and role conflict have been demonstrated by numerous studies. For example, Shirom (1989) found a positive correlation between role conflict and coronary heart disease. OSH professionals, with Occupational Health professionals, need to play a role in reviewing and analysing statistics on stress related illness in the organisation, such as coronary heart disease, irritable bowel syndrome, ulcers or depression and making senior management aware of the costs to the organisation.

6.45 Clarifying roles and responsibilities via job descriptions, especially when job/roles have altered, can result in a significant reduction in an employee's stress levels because work objectives and the expectations of co-workers are clearly defined. An effective performance-management process should enable employees to raise concerns about any uncertainties or conflicts they have in their role and responsibilities. Unfortunately, the performance management process, as applied, does not always encourage open and honest dialogue between employees and managers. This is a major area for OSH professionals to address because it inhibits important feedback on the effectiveness of processes and procedures that would otherwise require modification. Organisations can also clarify roles and responsibilities by ensuring that employees understand how their job/role fits into the overall aims and objectives of the organisation or department.

Job control and decision latitude

6.46 How well an employee performs at their job or role is the result of a multitude of organisational and personal factors. However, a key factor appears to be the degree of control they have over workflows, deadlines and demands. Nearly four decades ago, Hertzberg (1966) argued that employees are intrinsically motivated to achieve high levels of performance if job/roles are designed to increase personal or team responsibility and encourage autonomy and problem solving. Research evidence with regards to stress now shows that when employees have little decision-making latitude and when work demands are high, imposed inflexibly and without consultation, this is likely to lead to major negative health effects. For example, Karasek and Theorell (1990), in a wide-ranging landmark study, concluded that such working conditions resulted in up to three times the risk of heart attack in professionals and executives who had equally stressful jobs but had some degree of control over their work. Having little control over one's work is also linked to poor work performance and an increase in the likelihood of accidents.

6.47 The increased involvement of employees in setting work targets via mechanisms such as goal setting within performance management and the delegation of responsibility for solving and completing work tasks, perhaps via the development of work teams, will almost inevitably encourage employees to take greater responsibility for their actions, improve job satisfaction and reduce the likelihood of errors due to misunderstanding.

6.48 Increasing autonomy in how job/roles are carried out can however, also lead to greater stress. Therefore it is important that adequate learning and development programmes and other support mechanisms accompany any increase in responsibility.

Workplace support and relationships

6.49 Humans are social beings, we are constantly socialising and interacting with others. However, relationships can be a major source of stress as well as support. It is not surprising therefore, that the quality of relationships at work can have a significant impact on wellbeing, work performance and the likelihood of stress related errors and accidents.

6.50 Poor work relationships are often the result of interactions with colleagues and line managers that are characterised by a lack of trust, little support, and low interest in listening and attempting to tackle workplace problems. More direct causes of stress include abrasive personalities, authoritarian leadership styles, group pressure and workplace bullying. Workplace bullying can be very subtle and implicit, yet it is commonly recognised as being extremely distressing to victims.

6.51 Line mangers can help to support employees by setting achievable goals in consultation with employees and providing regular and constructive feedback as to how work tasks are being undertaken. Managers can also provide support by giving encouragement and praise for good work and by ensuring that employees have the necessary resources and competencies to carry out their job/role. At an organisational level, there should be systems in place to respond to individual concerns such as bullying, the use of authority without justification and unhelpful criticism.

Employee assistance programmes

6.52 The occupational stressors discussed above (**para 6.37**) are conceptualised as resulting from organisational systems that could be better designed to reduce the negative effects of stress. However, organisations more commonly attempt to lower the impact of workplace stress by providing employees with employee-assistance programmes (EAPs) such as telephone helplines, stress management training, counselling and health promotion.

6.53   The evidence for the effectiveness of EAPs is mixed and generally points to improvements in mental health but little or no improvement in job satisfaction, productivity or accidents rates. Reynolds et al (1993) found that stress management training lowered self-reported stress indicators but did not increase job satisfaction. Work counselling has been shown to reduce absenteeism and mental health symptoms and also shows some increase in job satisfaction (Cooper and Sadri, 1991). Similarly, health-promotion programmes that encourage employees to eat more healthily at work and attend health and fitness programmes tend to have few measurable organisational benefits. Worker health may increase significantly but much of this may be short-term if individuals do not make a commitment to maintain their new healthy lifestyle.

The risk management approach to occupational stress

6.54   Occupational Health and Safety professionals need to adopt a risk management approach to reducing the negative effects of occupational stress, specifically by tackling the underlying causes of stress. That is, to identify stressors as occupational risk factors and redesign job/roles and the work environment to make them more efficient and, therefore, less stressful. Arnold et al (1998) argue that 'employee assistance programmes present a high-profile means by which organisations can be seen to be doing something about stress'. They argued that EAPs merely assist employees to *cope* with the effects of inherently stressful work practices. In other words, they tackle the symptoms rather than the causes of work related stress.

6.55   The use of EAPs has increased significantly in recent years as organisations have recognised the negative effects of stress on their workforce. Such programmes provide a reasonable precaution to safeguard employee health and are regarded as easier to implement and less expensive than engaging in organisational development. The evidence, however, is that the cost of *not* investing in the working environment is much greater in the long run. Inefficient systems and procedures that result in occupational stress are very costly as they inevitably result in greater absenteeism, greater employee turnover and the resultant cost of recruitment and training, losing experienced and valued employees to competitors and increased accident rates. The 'hidden' costs include the effects of low morale, lack of commitment and resistance to change on client/customer relations and on the overall image and status of the organisation. A risk management approach to occupational stress is to view stress as an organisational risk factor that necessitates better management. OSH professionals can play an important role in initiating activities to change the culture of the organisation so that frontline and middle managers are made aware of their responsibility for identifying the causes of stress and to work with employees to remove or minimise them.

JOB/ROLE DESIGN

6.56   As knowledge and understanding of the factors that lead to organisational failure and loss have developed, so it has become increasingly clear that a vital component in creating a positive risk management culture is to design job/roles that mitigate against the effects of human error. In a practical sense, this is largely about designing job/roles that reduce stress and increase job satisfaction. Inefficient systems and procedures are likely to cause stress and therefore error, as are boring, repetitive jobs that do not utilise the human potential to work together to achieve goals. This element is particularly relevant to risk management as it is intimately linked to organisational development, ie putting in place the organisational structure and processes that assist organisations to succeed by focusing on reducing the potential for loss. Although job/role design is a vital component in the management of risk, it must be supported by other organisational factors such as effective responsibility and communication frameworks and OSH functions such as learning and development, performance management systems and the development of work teams.

6.57   Until the latter part of the twentieth century, jobs became increasingly monotonous and controlled. Many jobs were designed to minimise skill requirements, minimise the time and mental effort to perform a task and maximise management control. Jobs designed in this way had a human cost in terms of poor job satisfaction and stress leading to poor mental and physical health. With the steady decline of manufacturing industries in the UK and the related disappearance of many manual and semi-skilled jobs, such work practices and thinking has diminished. The exponential increase in the numbers of knowledge workers has necessarily resulted in the evolution of different work practices that better utilise employees' skills and talents. However, the influence of antiquated ideologies, particularly with regard to management control, still persist in some quarters and need to be challenged if organisations are to effectively manage the risks associated with poor job/role design. Attempts have increasingly been made to redesign work to enable employees to use their skills and talents effectively and to improve the quality of working life; this usually involves increasing one or more of the following job/role characteristics:

- Variety of tasks or skills – increasing use of capabilities.
- Autonomy – more control over when and how jobs are done.
- Completeness – whether the job produces an identifiable end result that makes the task more significant and meaningful for employees.

- Feedback from the job – improved knowledge of the results of work activities.

6.58 Other characteristics that are believed to be important include the quality of social interactions with colleagues and line mangers, and job/roles that require some element of problem solving and the implementing of solutions.

The growth of work teams

6.59 As the economic environment has become increasingly competitive, fuelled by new communication technologies and consumer demand, so work teams have evolved as one of the most effective and flexible mechanisms to achieve organisational goals. Where individuals or small groups worked on limited tasks that were closely monitored, now work teams engage in bigger projects and are responsible for identifying, developing and implementing solutions to a much greater extent than in the past.

6.60 Work teams are effective because they contain many of the elements identified as increasing job satisfaction, for example greater task variety and autonomy, and an increased sense of ownership of tasks. Work teams bring together the skills, experiences and insights of several people so they tend to outperform individuals or larger groupings without a specific function. High-performance teams invest much time and effort exploring, shaping and agreeing on a purpose that they feel belongs to them. Such teams are characterised by a deep sense of commitment to their growth and success as a team.

6.61 One of the most important aspects of work teams is that they more easily allow organisations to foster attitude change towards a new culture; this is because the internal dynamics of teams promote the development and adherence of group norms. Learning and development programmes for teams should therefore not only consist of the competencies required to carry out tasks and the interpersonal skills necessary to operate as a team, but also provide a mechanism for organisations to effectively promote the aims, attitudes and values they seek to create. The development of a risk aware culture is best fostered by encouraging teams at all levels to be aware of the organisational factors that expose the organisation to unnecessary risk.

6.62 However, work teams are complex systems to manage because they involve greater interaction between members in an environment that is less clearly defined. Work teams are made up of individuals with their own attitudes, values, beliefs and agendas, therefore the effective management of work teams requires the selection, integration and shaping of these personal elements to achieve organisational goals. OSH professionals have a large part to play in creating and developing such teams.

Improving the effectiveness of work teams

6.63   At a strategic level, human capital management is concerned with managing employees to meet business objectives such as growth and profitability. Job/roles designed around work teams with specific responsibilities generally prove to be most effective in this regard. Although work teams are not appropriate for every type of job/role, organisations that think creatively about restructuring work processes to include work teams have found them very successful once teams have formed into cohesive groups. Organisations that see the potential benefits of developing work teams should consider a number of factors that can assist in their improvement:

- Select people who will fit the culture of the organisation and the team but are still capable of independent thought.
- Make use of personality inventories to identify personality types and ensure teams are made up of a mix of the appropriate personality types.
- Take particular care over appointing and training/developing team leaders, emphasising that their role is to work alongside team members to achieve results by collective effort, not to 'stand outside the team' and deliver instructions from the 'sidelines'!
- Emphasise constructive team work as a key core value of the organisation.
- Get work teams to determine and agree with their team leaders their own objective standards, eg standards associated with meetings, scope of projects, level of supervision etc.
- Set overlapping or interlocking objectives for members who have to work together.
- Assess each member's performance not only by the results achieved but also on the degree to which they are effective team members.
- Clamp down on unproductive politics and confront disagreements openly and in a spirit of constructive feedback.
- Describe and think of the organisation as interlocking teams united by a common purpose.
- Devise and implement commitment and communication strategies that develop mutuality and team identity by developing cross-functional teams.
- Recognise and reward people who have worked well in teams.
- Introduce team rewards for achieving project objectives and commercial targets.

6.64   In order to be effective, work teams need a clear outline of their role and function, followed by a systematic approach to achieving their goals, ie identify needs, plan actions, decide responsibilities for actions, identify and obtain resources, carry out actions within agreed time frames, managers to provide effective support and feedback during tasks as required and a review of progress.

Limitations of teams

6.65 Of particular importance to the management of risk is the dangerous phenomenon known as 'groupthink'. Groupthink occurs when a group's need for unanimous agreement prevails over the need for careful consideration of the pros and cons of an argument and alternative possible actions. The need to maintain harmony and morale of the team supersedes the raison d'être of the team, that is, to meet organisational goals. Groupthink results in shutting out alternative views, exaggerating supporting evidence and suppressing dissent within the team. Janis (1972) points out that in the decisions preceding many disastrous business and political decisions, the risk of negative consequences could and should have been anticipated but were blocked by groupthink. The most recent case has been the global 'groupthink' about intelligence on weapons of mass destruction in Iraq, prior to the 2003 invasion. The term 'groupthink' was specifically mentioned in subsequent reports on intelligence failures.

6.66 The specific symptoms of groupthink are:

● Illusion of invulnerability – teams underestimate the likelihood of being wrong, especially if they have been successful in the past.

● Stereotyping – teams are quick to classify 'enemies', and do not notice discordant evidence.

● Pressure to conform – team pressure is put on dissenters.

● Pluralistic ignorance – because team pressure leads to no objections, this is seen as unanimous agreement.

● 'Mind guards' – canvass team members for support and keep bad or contrary news from leaders.

6.67 Autocratic leadership styles tend to encourage groupthink, and therefore a more democratic approach that encourages discussion and criticism is recommended. Simply being aware of the tendency to groupthink will allow team leaders and members to be cautious of such team dynamics and perhaps allow for one team member in rotation to play 'a devil's advocate role', specifically to make sure that groupthink does not happen.

COMMUNICATION FRAMEWORKS

6.68 Ensuring that communication systems are effective is a challenge for most organisations but the issue becomes particularly relevant when related to the management of risk. Effective communication is the foundation on which risk management systems are based. Embedding a culture of risk awareness requires risk to be managed in an organisation-wide, integrated manner. This process necessarily requires a high level of communication efficiency.

6.69 Occupational and Health professionals can support the development of a communications plan, because of their organisation-wide remit. OHS professionals who are operating at a strategic level should also be involved in implementing the risk management strategy and be able to advise on how communication issues can impact on risk management and organisational effectiveness more generally. Effective risk management requires an effective communications framework. It is essential that OSH professionals ensure that the development of communication and dialogue is an integral part of the organisational strategy and for effective role models to be recognised, rewarded and celebrated. In addition, it will be necessary to invest in resources such as up-to-date communication technologies, learning and development programmes, the appropriate use of rewards and investment in time and resources in developing an effective communications framework. Such an investment is of course, also a key building block in the planned development of an organisation. Therefore the benefits of using the risk management process to build a framework for communication will benefit the organisation in ways that go beyond the management of risk.

6.70   Because of the critical nature of communication, it is useful to think of communications as a specific management process in its own right, rather than undertaken on an ad hoc basis to facilitate the development of other processes. When a framework for communication is developed, the following should be considered:

- It incorporates a common vocabulary.
- There are clear roles and responsibilities for ensuring that communication is adequate, appropriate and timely.
- It includes a list of all stakeholders and the information they require.
- Key elements of information to be distributed to stakeholders are identified, including when it is disseminated and received.
- A differential system of communication content and processes is in place, to ensure that only timely and relevant information is provided to any one group/person, so as not to create 'information overload'. Information overload is sometimes worse than no communication at all. So avoid communicating unnecessary information as this is likely to render the whole communication process ineffective.
- Specific methods to communicate certain information, eg email, newsletter, written reports, workshops, etc, and including the repetition of key information using different mediums.
- Encouraging individuals and teams to take responsibility for accurate communication, including confirmation, clarification and repetition.
- Including communication issues and the use of a communication framework in any risk management learning and development programme.
- Setting up an independent audit to check that communication frameworks are in place, adequate and in use.
- Ensuring that responsibility for communication is included in people's objectives and that effective rewards and sanctions are applied.

6.71   Actions and decisions on risk management require multi-directional information flows between organisational levels and between and within functional areas. Information also needs to flow between the organisation and related external bodies. A focus on developing communication systems will ensure that risk management strategies, goals and objectives are communicated and the methods, reasons and timings are understood and agreed by all involved in the process.

6.72   Clear top-down channels of communication serve to communicate the purpose of risk management policies and the values and beliefs on which they are based. They also serve to reiterate the commitment of senior managers to its implementation and continuous development. Inadequate communication can result in risk management projects no longer supporting the business direction or losing impetus.

6.73   Information that flows upwards from lower levels provides vital feedback to senior managers as to the organisation's risk status and progress in implementing the risk management system. Therefore, this will influence senior-level strategic planning and decision-making. Inadequate feedback from lower levels can lead to unrealistic expectations from senior managers and to a disruption or breakdown in the implementation of the system. An important aspect of establishing clear channels of communication that flow upwards from operational levels is that it creates involvement in the process of risk management at all levels and helps to establish a culture of risk awareness and control.

6.74   The importance of effective communication extends to communicating with external stakeholders, such as suppliers, customers and investors. To ensure that costly misunderstandings are not created, communication with external stakeholders should be brought under the same communications framework as that operating within the organisation.

6.75   It can be argued that the reputation of an organisation depends to a large extent on what and how it communicates, therefore everything an organisation communicates (or does not communicate) in writing, verbally or via other mediums will affect how it is perceived. Mechanisms should be implemented to ensure that the risk of damaging communications leading to negative perceptions are minimised.

6.76   There is increasing pressure on organisations to communicate more fully and openly to stakeholders external to the organisation, for example to shareholders via the annual report. High-profile cases such as the collapse of Enron and WorldCom have resulted in the loss of confidence in corporate reporting and demands for evidence that the organisation has in place effective risk management programmes, particularly those that take into account non-financial risks such as organisational factors that are more difficult to quantify but are often the cause of failure. As has been discussed, organisational factors include communication and responsibility frameworks, job/role redesign to minimise stress and error etc. In addition, there are likely to be new legislative requirements placed on companies to report more fully. In **Chapter 7**, recent developments on a UK initiative entitled 'Operating and Financial Review' are discussed.

6.77 Effective use of external information and external communications is becoming increasingly important as organisations become aware that the business environment in which they operate is ever more sensitive to factors originating outside the organisation. The increasing interconnectedness of the marketplace, as a result of integrated global economies, communication systems and the media has brought huge business opportunities. However, socio-economic factors, market forces and the activities of external pressure groups can no longer be ignored if organisations are to survive and thrive. Consequently, organisations can benefit in many ways by adopting the principles and practices of 'corporate social responsibility'. This is also discussed in **Chapter 7.**

6.78 A globally interconnected infrastructure also means that organisations are more vulnerable to distant events – economic crises on the other side of the world, catastrophic events such as those on 11 September 2001, virus attacks on IT networks, or competition from cheaper foreign goods and services. As a result of the above factors, organisations cannot afford to ignore external information and should seek out and engage with sources of information that may indicate how external factors can adversely or positively impact on their success, or perhaps even survival. The ability for organisations to examine and manage external risks as an integral part of the risk management process will enhance organisational success.

6.79 The development of a common vocabulary of risk (see **Chapter 1**) and a framework for communicating risk related issues should assist in evolving a culture of risk management and risk awareness, and also help to protect against human error that results from misunderstanding and miscommunication.

6.80 As discussed in other chapters, issues relating to risk have traditionally developed independently in different functional areas, ignoring the interconnected nature of risk and the possibility of learning from good practice in other areas. The systems necessary to manage risk are, however, the same across the organisation as risks have interrelated and accumulative consequences across functional areas. The development of a common language of risk is important in facilitating risk communications as different functional areas often give different meanings to key terms. A common vocabulary of risk will allow for a clearer identification of interconnected risks and, in addition, promotes better relationships between functional areas and improved problem solving. In **Chapter 1** we have included some examples of the definitions from two risk management standards that can be used to develop a new vocabulary or review an existing one.

LEARNING AND DEVELOPMENT AND THE MANAGEMENT OF RISK

6.81   Risk management provides an effective mechanism for organisational growth and improvement. This is because uncontrolled risks result from inefficiencies that ultimately lead to an increase in error, loss and accidents. Implementing a risk management system, as with any organisational change process, will require learning and development to provide employees with the necessary skills, knowledge and attitudes to achieve the organisation's new strategic goals. The lack of appropriate learning and development is a fundamental cause of loss and accidents in the workplace therefore learning and development programme initiated and implemented by OSH professionals are seen as a key component in establishing a positive risk culture. Implementing a risk management system will require employees at all levels and functions within an organisation to alter their behaviour and attitudes towards risk. The abstract nature of risk will also require changes in the perception of risk and the values that guide risk related behaviour. Therefore a coherent learning and development plan is necessary to achieve this.

6.82   A strategic OSH risk management approach to learning and development should clearly show how such activities will contribute to the management of risk. As discussed, risk management goals could include minimising organisational failure and loss, enhancing performance or better controlling the uncertainties associated with new business development. Learning and development should therefore be considered an investment in human capital and not a cost. However, in order to justify the expense of these activities, they must be effective, and seen to be effective. Learning and development programmes that do not incorporate the intricacies and integrated nature of risk management are likely to cost less but not be very effective. Effective programmes necessarily view risk management as an organisation-wide, integrated process.

6.83   Because of the organisation-wide nature of risk management, learning and development plans should be part of, and not separate from other OSH strategies such as including performance on Health and Safety issues in personal development programmes and reward strategies.

6.84   Learning and development cannot, however, create organisational change independently. Clear risk management policies are essential, as is a commitment from senior managers to implement risk management procedures. As discussed above, it is also important to develop organisational factors such as risk accountability frameworks, communication frameworks and job/role redesign; specifically, job/roles designed to minimise error and stress and increase job satisfaction.

6.85   At a fundamental level, any risk-focused learning and development programme is likely to operate on three levels:

1   Organisation-wide level.
2   Management level.
3   Job/role or task level.

6.86   The following processes form part of any learning and development programme, but their focus will vary depending on the organisational level at which they are being applied:

● The identification of risk.
● The assessment of risk.
● The development of skills, competencies and knowledge to make effective use of risk-control systems.
● Monitoring the effectiveness of learning and development programmes.

Developing knowledge and understanding of risk at the organisation-wide level

6.87   Organisational risk is not function specific and often involves an inter-related set of circumstances across departments or organisational activities. Therefore, implementing a risk management strategy will initially require an organisation-wide programme to raise awareness of the nature of organisational risk and the benefits of controlling risk. This should include the creation of a common risk vocabulary to facilitate communication as different functional areas often have different terms to express similar issues. Ideally, all managers and employees should undergo such awareness learning and development programmes. Managers and employees will also require specific knowledge and an understanding of risk management policies and practices depending on their role or level within the organisation.

6.88   Occupational Health and Safety professionals need to encourage the development and understanding of organisation-wide risk factors. This is best achieved by involving employees and managers in a structured process to discuss and analyse specific and relevant risk issues. Employees have a wealth of knowledge and experience of working in the organisation and should be encouraged to share this with others. Cross-functional and cross-level discussion groups will promote a greater understanding of how different organisational functions and levels of responsibility can impact on each other. This is necessary as latent failures are often due to organisational factors that are cross-functional, such as communication failures or unclear accountability for processes that continue across functional areas or that connect functional areas together.

6.89   The key objective of awareness programmes is to change attitudes and values about risk management by actively involving employees – and ideally, other related organisations such as those in the supply chain. Involvement results in ownership and therefore greater commitment to the risk management process. It also enables the organisation to benefit from valuable expertise and 'local' knowledge that may otherwise be lost in a more top-down approach to training.

6.90   There are a number of tools and techniques available for communicating a corporate 'message' across the whole organisation. In our experience, a process entitled the OPC Process® has proved particularly effective in communicating the organisational and personal benefits of behaving in a risk aware manner and in developing a sustainable risk aware culture.

6.91   The OPC Process® is designed to communicate and engage large numbers of employees with any key corporate 'message'. It has been used successfully in many types of organisation, private and public, and in many countries. In the context of risk management, it is ideally suited for use by large organisations that wish to implement a risk management system as part of their programme for organisational development.

6.92   The process is successful because it combines a sound understanding of how attitudes and behaviour can be changed to meet organisational needs with the effective translation of the organisation's 'message' into engaging workshops that internalise the 'message' for each participant.

6.93   A logical series of interactive and engaging tasks guide cross-functional teams to analyse and discuss specific risk management issues. The process and materials are structured in such a way as to allow managers or employees without specific risk related knowledge to act as facilitators. The team nature of the process means that a large number of managers/employees can be made aware of complex and interrelated risk management issues in a focused and timely manner. Involvement and consultation also results in attitude change towards risk, and the identification of risk related behaviours. It also results in a clearer understanding of how systems can be modified to control risk and provides the foundation on which to carry out organisational development.

6.94   In **Chapter 7**, a tool called 'RiskFrisk® – OSH Risk Management' is described. The tool can be used to identify the risk profile of an organisation and therefore assist organisations in identifying and understanding where risk management systems need to be enhanced, and where risk taking behaviours need to be replaced by intelligent risk taking. The strategic risk goals identified as a result of a RiskFrisk® can be translated by the OPC Process® into a set of key messages to be communicated across the organisation and internalised by employees.

Developing risk related competencies at management level

6.95   A number of factors will affect the type of learning and development necessary to imbed an integrated risk management system. Many of these factors have already been discussed but are summarised here from the perspective of learning and development for managers. Clear lines of responsibility and accountability for the management of risk should be established and the competencies necessary to carry out such responsibilities acquired through learning and development. A risk management competency framework for specific functions can be arrived at from an analysis of what risks managers are required to manage. A risk-profiling tool such as RiskFrisk® will identify specific functional risks and the behaviours and processes necessary to control such risks.

6.96   At the management level, the identification and evaluation of risk should form part of every learning and development programme related to the function for which managers are responsible. It should be imbedded into competencies that relate to actual behaviour in a specific setting rather than being bolted on as an independent unit of risk management.

6.97   It is necessary to monitor how effectively managers carry out tasks associated with competencies. Performance management systems that include risk management will ensure that it remains high on the agenda for managers. Reward systems that recognise effective risk management and intelligent risk taking will also make a significant contribution in encouraging managers to behave in accordance with risk management policy.

6.98   Managers should be aware of their powerful influence as role models and seek to create a positive risk culture by maintaining a visible focus on the management of risk. This is often referred to as 'Walking the talk'. For example, managers should never allow risk taking practices to be used arbitrarily under certain conditions, but not others.

6.99   Managers sometimes find it difficult to discipline employees for engaging in risky behaviour for fear of the effect it would have on their working relationship. In this way small offences become common practices and greater offences develop. Managers can be assisted in their role by unambiguous and transparent policies for deliberate transgressions of procedures likely to lead to error or loss. Managers can also benefit from learning and development programmes in how to give praise for intelligent risk taking and for enforcing risk aware behaviour in a non-threatening but assertive manner.

6.100   If there is a clear and unambiguous commitment from senior managers to organisational development via risk management, there should be no conflict of interest for middle managers between their role in meeting 'production' targets and their role in managing risk. This should allow managers to seek to manage and improve organisational factors that can mitigate against risk taking behaviour, eg poor communication systems and job/roles that result in job dissatisfaction and stress. OSH professionals operating at a strategic level should have the necessary expertise to assist managers in developing such organisational processes, especially when these processes are cross-functional and involve people management issues – for example, developing communication and accountability frameworks.

6.101   In order to help to maintain and improve the risk management process, it is important that managers are supported in this. A cross-functional forum to discuss issues relating to the management of risk and the exchange of good practice is therefore considered vital. Such a forum should include risk management and OSH professionals, as they would be able to provide a broader, organisation-wide perspective. A valuable function of such a forum would be to co-ordinate and analyse the progress of risk management so that senior managers could be kept informed. The forum would also provide a mechanism for senior managers to communicate strategic risk management goals to those charged with implementing them.

Learning and development at the job/role or task level

6.102   Research shows that, at this level, risk taking behaviour for routine tasks is largely habitual and occurs without conscious thought. The implications of this are discussed in greater detail in the section entitled 'The perception of risk', but the key point with regards to learning and development is that it is very difficult to change risk related behaviour for routine tasks by simply attempting to change attitudes towards the perception of risk. An effective risk management learning and development programme for routine job/roles and tasks will require the design and implementation of specific procedures that minimise risk related behaviour in the first instance. Learning and development for new procedures is then best focused on behavioural change related to specific work situations.

6.103   Some key principles include:

- Behavioural change is best achieved via demonstration and practice of positive risk related behaviours rather than simply through instruction.
- It is also necessary to alter perceptions as to the negative consequences associated with deliberate risk violations. This, however, is more effective for new employees than for those who have engaged in such behaviour without negative consequence in the past. Innovative communication methods that engage employees in discussing their perception of risk can be effective (see the section on 'The perception of risk').
- Learning and development should make good use of the organisation's historical background on risk control. Operational feedback systems that collate data on human error, accidents and 'near misses' provide a valuable sources of information that will place learning and development within the context of the organisation and specific functional areas.

- Learning and development should result in more than just being able to carry out or recite a series of complicated steps. It should provide the background contextual knowledge that allows individuals to evaluate the likely outcomes of their actions.
- There should be regular reminders as to what constitutes a positive risk culture.

INTRODUCTION TO HUMAN FACTORS

6.104   The term 'human factors' is increasingly being used in the organisational risk and safety literature. It is a useful concept because it draws attention to how people interact with organisational systems, procedures, tasks and equipment.

6.105   The term 'human factors' can sometimes include ergonomics, which is specifically concerned with matching human physical and spatial abilities with workplace practices to minimise physical fatigue, strain or injury. It is also sometimes used to explain how organisational factors such as job/role design influences human behaviour. For the purposes of this chapter, a distinction is made between 'organisational factors', eg job/role design, communication and responsibility frameworks, etc, and 'human factors' that essentially describe intrinsic human characteristics such as cognitive abilities and attitudes that impact on human error and risk taking behaviour. This perspective is derived from the work of organisational psychologists and utilises psychological theory and research methodologies to understand workplace error and risk. Human factors include cognitive abilities such as attention span, short-term memory capabilities, and problem-solving strategies, however almost any human characteristic will influence risk related behaviour. Personality differences, cognitive style, attitudes, values and beliefs are just some of the categories that could be examined with regards to risk related behaviour.

6.106   OSH professionals supporting the design and implementation of organisational factors such as responsibility allocation or learning and development programmes need to be aware of how human factors can impact on the management of risk. OSH is ideally placed to provide input on these matters, as are operations management and Human Resources management.

6.107   Almost any human characteristic will influence risk related behaviour. Personality differences, cognitive style, attitudes, values and beliefs are just some of the categories that could be examined with regard to risk related behaviour. The human factors discussed in this chapter are however, considered to be particularly pertinent because they appear to be fairly generalised and consistent in their influence on risk. They are:

- Perception of risk.
- Motivation and risk related behaviour.

- The effects of stress as influenced by organisational factors.
- The nature of human error
- Classifying and limiting human error.

THE PERCEPTION OF RISK

6.108 The perception of risk is a human factor because, by definition, an individual's perception is not based on quantifiable evidence but on aspects such as previous experience, values, attitudes and personality characteristics. How risky a situation is perceived to be will, therefore, be different for different people. People tend to disagree about what is risky, dependent on how much control they feel they have over a situation or whether they have experienced any negative consequences in the past. Thus, experienced car drivers may feel less at risk driving their own vehicles than travelling by aeroplane, even if a more objective analysis of the risk of an accident would suggest otherwise.

6.109 Research into the perception of risk has attempted to identify and explain the factors associated with the perception of risk in the workplace and why people behave as they do when faced with situations that could have potentially negative consequences to themselves, to others and to the organisation.

6.110 One might assume that before engaging in a potentially hazardous activity, an individual would make a conscious calculation as to how risky the situation actually is. For example, a decision to take a risk and bypass an established procedure because it is cumbersome or time-consuming would be weighed against the experience of having taken this risk in the past without negative consequences, and against the benefits of completing the task more easily or quickly. Research in this area suggests that, in fact, such conscious and rational decision making rarely occurs, especially for routine tasks. It appears that much risk taking behaviour in the workplace occurs out of habit, a habit that is shaped by a reality in which negative consequences rarely happen.

6.111 There are a number of implications that can be drawn from the above analysis. Firstly, if risk taking behaviour, particularly for routine tasks, occurs without much rational consideration of risk, then risk taking cannot be attributed to a miscalculation or a misunderstanding of the risks involved. Rasmussen (1983) argues that, for this reason, when negative consequences eventually occur, blame should reside not with the individual but with senior managers whose role it is to consciously and rationally consider the possible causes of risk taking behaviour and implement procedures and practices that minimise such incidents, particularly for routine tasks.

6.112 A second implication of the fact that work tasks are carried out largely without consideration of the risks involved is that it is difficult to cause individuals to consider the actual risks associated with their work and alter their perception of the risks involved. For this reason, despite millions of pounds spent on campaigns to make people more 'safety conscious' at work, the effects of such campaigns have only proved marginally effective, especially when introduced into areas where the 'risk taker' is already familiar with the situation and has decided the level of risk they wish to attribute to a particular task.

6.113 The extent to which risk is perceived to be associated with an action is dependent on a number of factors, including:

- The individual's perceived level of skill on a task.
- Past experience of negative consequences.
- The probability of detection and the types of sanctions applied for violating safety rules.
- Rewards for behaving safely such as praise or promotion.
- Rewards for violating safety rules such as more free time, kudos from less experienced employees.
- Personality characteristics – research suggests that people vary as to how much risk they are prepared to take.
- Age – research shows that risk taking is greater amongst individuals aged under 35, especially for males.

6.114 The 'risk-homeostasis' theory (Wilde, 1982) suggests that people make use of the above factors to identify and maintain a personal level of perceived risk for a given task. Once this has been established, little conscious thought goes into re-examining the situation unless circumstances change – sometimes catastrophically. The risk-homeostasis theory postulates that if a task is made less hazardous by, for example, the use of better-designed equipment or less hazardous procedures, the person will continue to be motivated to maintain the same personal level of perceived risk as that previously established for the task. In effect, the individual takes greater risks to counter the effects of a more controlled working environment. To use a car analogy, as cars have become easier to handle and safer to drive, so speeds have increased and behaviour that would have been previously considered risky is now no longer thought of in that way. Indeed there is some evidence that, as cars have become safer, people take more risks, thinking that the safety features in the car will protect them or prevent negative consequences. For example, anti-lock brakes may have encouraged people to drive faster, thinking they can more easily avoid an accident. In reality, 95% of vehicle accidents are still caused by human error.

6.115 A second theory that can shed light on the perception of risk is the 'zero-risk' theory (Naatanen and Summak, 1976). It is related to the risk-homeostasis theory but suggests that individuals are motivated to seek situations in which risk is perceived to be minimal. In reality there may be considerable risk but this is not rationally acknowledged until safety rules are breached. Research shows that individuals are likely to operate at the outer limits of risk margins and to keep this margin as small as possible. To continue the driver analogy, the driver will only begin to consider the risk of such an activity if they are driving beyond the legal speed limit. The implication of this theory is that wide safety margins should be enforced to reduce the consequences of potential accidents and that the perception of risk results from the inculcation of habits brought about by the design of systems and procedures. This theory is clearly relevant to systems and procedures for managing risk at all levels and in all areas of an organisation.

6.116 The management of risk is therefore an ongoing process and simply attempting to change attitudes towards risk is not very effective. A much more effective approach is to implement specific procedures and practices that cause individuals to behave in less risky ways. For example, in order to cause drivers to drive more safely, speed restrictions, traffic calming systems or speed cameras are more successful than advertising campaigns about the dangers of speeding.

6.117 In addition to restructuring work practices so that they result in less risky behaviour, the perception of risk can be altered by increasing the benefits associated with behaving in an intelligent risk taking manner. For example, management giving praise for working safely has been shown to be a strong motivator to comply with safety procedures. Publicly recognising and celebrating safe workers and safe managers also helps to alter the perception of risk taking as something that is undesirable within the organisation. Aligning intelligent risk taking behaviour with performance appraisal reviews and promotion is perhaps the most effective way to use positive reinforcement as it is directly related to the individual's risk taking behaviour and is continuously monitored.

6.118 The perception of risk can also be influenced by increasing the negative consequences of violating risk management systems, for example by increasing the probability of detection via better monitoring and by creating a system of transparent and consistently applied disciplinary procedures with regards to dangerous or potentially harmful risk taking.

6.119 It is vital that effective learning and development programmes are used to ensure that employees know how to use equipment safely, know how internal controls on key business and operational processes need to be applied, and know how to carry out procedures so that the risk of error or accidents is minimised. However, it is also necessary to alter perceptions as to the negative consequences associated with deliberate violations of risk management procedures. As has been discussed above (**para 6.112**), this per se, is of limited effectiveness particularly for employees who have engaged in such behaviour without negative consequence in the past. This type of learning is more effective for new employees and should focus on the potential consequences of injury on family and career, effects on work colleagues, the extent of direct financial loss to both the individual and to the company, and the effects of less tangible losses such as damage to the reputation of the organisation.

6.120   Some ways in which OSH professionals can promote intelligent risk taking behaviour are:

- Help to redesign job/roles, procedures so that barriers are placed against negative risk taking eg signing-off or acknowledgement procedures. (see section on Classifying and limiting human error, commencing **para 6.142**)
- Instituting wider safety margins
- Effective recruitment and selection processes where the organisation's focus on risk management can be described and discussed.
- Induction and initial training on risk issues, where details can be provided about the organisation's vision, values and strategies with regards to risk management and internal control and how this is translated into individual responsibilities.
- Subsequent internal learning and development where appropriate elements about risk management processes can be included on a regular basis.
- Introduce a strategic review and ongoing processes for job/role redesign and team development.
- Ensure that job descriptions contain specific responsibilities for risk management.
- Ensure that performance-management systems include a risk management element.
- Ensure that sanctions and rewards systems include performance measures related to risk management.
- Establish systems to recognise good performance in risk management compliance and intelligent risk taking behaviour.
- Ensure that Human Resources systems operate in accordance with risk management principles.
- Ensure that disciplinary procedures include examples of unacceptable risk taking behaviour.

MOTIVATION AND RISK RELATED BEHAVIOUR

6.121   Motivation is central to most human activities because it governs our choice of behaviour and attitudes. In order to understand how people behave with regard to taking risks in an organisational environment, it is necessary to understand what motivates such behaviour.

6.122   Motivation at work is a complex issue because it relates to a number of interrelated human and organisational factors. For example, human factors include individual differences in work competence, perceptions of the situation and self, differences in attitudes and values. Organisational factors include procedures, the design of job/roles and equipment, learning and development programmes, clarity of roles, accountability and communication frameworks, reward systems, etc.

6.123   A number of theories have been proposed to explain human motivation. Such theories have had considerable impact on attitudes towards work, job performance and the nature of work itself. No single theory can adequately prescribe how motivational factors can be utilised to manage employees. However, theories provide the best guide as to how motivational factors influence risk related behaviour in the workplace.

6.124   One of the main theories is *reinforcement theory* advanced by behavioural psychologists such as Skinner (1938). He proposed that any behaviour that is rewarded would be repeated. This is known as *positive reinforcement*. In the workplace, this could be any reward, praise, approval or money. However, it could also be greater responsibility, autonomy or access to decision making processes.

6.125   *Punishment* is regarded as any unpleasant occurrence following a particular behaviour. This is likely to deter the behaviour in future. In the workplace, a strong 'punisher' could be a reprimand from a manager for not following agreed procedures. However, a punisher could also be something less dramatic, such as not having a suggestion for an improvement in risk management taken seriously, which would deter future initiatives or suggestions to improve risk management.

6.126   *Negative reinforcement* is described as any behaviour that results in avoidance or escape from a negative consequence. For example, an employee may agree to take on an increase in workload, not because it is rewarding but because they fear being negatively appraised for not doing so.

6.127   According to this theory, in order to cause behavioural change in individuals so that they do not engage in potentially risky behaviour, it is necessary to consider that individuals will balance the value of rewards for behaving in a intelligent risk taking manner against the value of engaging in more risky behaviour. For example, getting the job done more quickly to increase leisure time, set against the probability of being caught or causing negative consequences. Individuals will be motivated to achieve the greatest reward; thus deliberate violations of known risk management systems can be understood as the result of a 'rational' calculation that the benefits of such a violation will outweigh the likelihood of a negative consequence occurring. Once a decision has been made concerning a particular behaviour, it no longer receives much conscious thought and becomes routine.

6.128   The difficulty for risk managers is that the certainty of short-term rewards is set against punishers that may be weak, or may never occur. Thus, risky work practices persist because they are rewarding. In addition, the rewards for working in an intelligent risk taking manner may be minimal. In some organisations, it is rare for anyone to get praise for working strictly to the rules.

6.129 Even if near-misses occur, the irony is that they may actually reinforce risk taking behaviour, as success at avoiding a negative consequence may be perceived as skilled behaviour. To counter this, it is necessary to associate competence with taking intelligent action, rather than risk taking action.

6.130 Reinforcement theory therefore suggests that, in order to redress the balance towards the avoidance of risk, there must be a relative increase in the benefits of adhering to risk management procedures as compared to the benefits of behaving in a risk taking manner. Management giving praise to others for working to minimise risks has been shown to be a strong motivator to comply with risk management practices. It is also the case that increased involvement and responsibility is, in itself, rewarding to most people, especially if it is associated with access to decision making and supported by adequate learning and development programmes. The implementation of an organisation-wide risk management process will encourage involvement and responsibility and thus motivation to alter behaviour.

6.131 Effective learning and development is necessary in order to highlight the negative consequences associated with deliberate violations of risk management procedures. For example, injury, effects on work colleagues, effects on family and career, financial losses both to the individual and to the organisation, and losses related to the reputation of the organisation, etc.

6.132 A related issue with regard to increasing negative consequences of risk taking behaviour is to raise the probability of detection by better monitoring of adherence to procedures and by creating transparent and consistently applied disciplinary procedures.

6.133 Social Learning Theory contends that individuals are motivated not only by reward and punishment but by observing others receiving rewards and punishment. Individuals are motivated to imitate the behaviour of others if they observe that they are being rewarded for their behaviour. Any behaviour is more likely to be imitated if the 'role model' is perceived to be successful, charismatic or of a higher status. The implication of this is that managers, particularly senior managers, must lead by example if intelligent risk-based behaviour is to be improved.

6.134 Equally, such 'role models' can seriously undermine the development of a positive risk management culture by showing less than total commitment to the process. Individuals perceived as high in status, although not necessarily of higher authority, will also act as role models. It is important therefore, that such individuals are identified and brought on board with regard to risk management. Giving high status to those who adopt a risk aware and risk management approach will also promote a positive risk management culture.

6.135   Although it might be expected that incentive programmes such as receiving bonuses for reduced accident rates or wastage rate due to fewer errors would reduce risk related behaviour, such reward schemes are, however, only marginally successful. This is because rewards are given for *not* getting injured or *not* making errors rather than being rewarded for specific risk management related behaviours. Likewise, campaigns exhorting adherence to risk management procedures are not generally successful because motivation in risk avoidance appears to be highly situation-specific. Our experience is that it is much more successful to reward good behaviours, ie rewarding a manager for undertaking actions that improve the management of risk, rather than rewarding them for reducing the level of losses or other negative consequences.

6.136   *Goal theory*, usually attributed to Locke (1968), attempts to focus reward systems and involve the individual in personal goal setting. The basic concept is that employees are motivated to achieve specifically stated goals whereas simply urging people to 'work safely' has little or no effect. Goals should also be perceived as personally achievable otherwise they are ignored. Goal theory has similarities to reinforcement theory in that motivation is the result of rewards for achieving goals. However, goal theory states that employees should be actively involved in negotiating their own goals, therefore reinforcement does not come only from external sources (ie praise from line manager, bonuses, etc) but also from the satisfaction of meeting one's own goals. The following guidelines have been found to be useful:

- Risk management goals should be challenging and SMART (Specific, Measurable, Achievable, Realistic and Time-framed).
- Support elements should be provided, ie resources as well as encouragement, moral support, etc.
- Feedback en route to the goals should be provided in order to sustain and increase motivation.
- Risk management goals should be negotiated with employees, as involvement will increase commitment and motivation.
- Individual performance on risk management activities should be integrated with performance management, rewards systems and succession planning processes.

THE NATURE OF HUMAN ERROR

6.137   The study of human error has revealed a number of pertinent issues that have important implications for risk and OSH professionals engaged in the management of risk. At an organisational level, human error is considered to be largely the result of organisational factors that have not been controlled and therefore do not mitigate against this risk. An understanding of the concept of human error in an organisational context and how it can be classified in a generic sense will assist in the design of processes and procedures that reduce the likelihood of error and its consequences.

6.138 Human error has been defined generally as 'the failure of a planned action to achieve a desired goal' (Reason, 1990). This definition implies that there was an intention to achieve the correct outcome, but that the goal was unattained. The human factors approach to error does not therefore presuppose that someone is to blame because human error is, by definition, unintended. In other words, it does not make logical sense to blame an individual for an unintended act. For example, in the case of the Ladbroke Grove rail disaster or the *Herald of Free Enterprise* disaster, it was the person who triggered the incident who got the public blame, yet further scrutiny revealed a catalogue of organisation-wide system failures that were created or ignored over many years. Even deliberate violations of procedures can be seen as 'perpetrators' not clearly understanding the consequences of their actions. Such risk taking behaviour is considered to be the result of ineffectively managed organisational factors such as poor learning and development programmes, ineffective procedures or the failure to enforce procedures.

6.139 It is a truism to say that 'to err is human' (A. Pope, 1711). Not only is error essential to learning but making mistakes is also a natural feature of human existence. Failure to acknowledge this in individuals or in an organisation, ie systems created by humans, is likely to lead to poor performance at best and disaster at worst because error feedback is not used to guide future behaviour.

6.140 One of the most enduring characteristics of human nature is the ability to solve problems. This involves utilising feedback from mistakes to continuously reinterpret the problem and eventually achieve the desired solution. Research has shown that 'trial and error' learning appears to be the most effective way to learn new tasks and, in fact, the propensity to make errors in order to learn is 'hard wired' into brain functioning as an evolved mechanism because it promoted the survival of our species. The importance of error feedback is recognised in safety-critical systems where simulations are used to allow operators to experience and therefore learn from the consequences of their actions in a safe environment. Simulated learning environments include the use of flight simulators in civil and military aviation, shutdown simulations in the nuclear power industry or practising emergency fire-exit drills.

6.141 Trial-and-error learning appears to be an enduring human characteristic. However, we are increasingly caused to interact with complex systems where the consequences of our actions may not be at all obvious. Therefore, although we may be intrinsically motivated to attempt new solutions to make procedures more efficient from our own personal perspective, the bending or breaking of established procedures may be entirely inappropriate in a modern working environment. It may also be true, however, that existing procedures are inappropriate given the nature of the task.

CLASSIFYING AND LIMITING HUMAN ERROR

6.142   It is difficult, if not impossible, to eradicate human error from organisational systems, although much can be done to reduce the risk of human error. An understanding of the types of errors employees are likely to make, and the circumstances in which they are likely to be made, provides a useful guide to designing organisational systems that seek to limit risk related behaviour.

6.143   The term 'human error' describes occasions where the intention to carry out a particular sequence of actions or thought processes does not result in the desired outcome. The intention to achieve desired outcomes may fail because actions did not go as planned (called a slip), or because the plan itself was inadequate (called a mistake). This provides the basis for the distinction between two different categories of human error.

Slips and lapses

6.144   Slips typically occur when carrying out familiar tasks that do not require much conscious thought. They usually result from a lack of attention. Stressful circumstances such as having too much to do, having to work too quickly or simply feeling unfairly judged by colleagues or line managers can result in cognitive overload, which makes slips more likely.

6.145   Lapses are similar to slips in that the intention is correct but error occurs because of a failure of memory. Lapses result in a failure to carry out actions at the appropriate time or lose a place when carrying out a procedure or task. Like slips, lapses can be exacerbated by stressful circumstances.

Avoiding slips and lapses

6.146   Slips and lapses are not usually the result of ineffective learning and development programmes yet they can sometimes lead to fatal errors. Workplace stressors leading to cognitive overload or stressful situations that originate outside the organisation in the home environment are likely to increase the probability of this type of human error at work. Slips and lapses can be minimised by designing procedures that are resistant to such errors. This is particularly important for safety-critical procedures or when operating machinery. It is also important when designing risk management systems to enhance internal controls.

6.147   The principles involved in reducing slips and lapses apply equally to carrying out procedures and operating machinery. They are presented here in some detail, but are not meant to be prescriptive, especially given the intricacies and pressures of business in the real world. Controls to minimise slips and lapses include:

- The use of checklists to ensure appropriate actions have been carried out or have been carried out in the correct order.

- Causing individuals to make some sort of positive acknowledgement at vital points in a procedure before proceeding further, for example, signing-off procedures.

- Procedures should be clearly and unambiguously named. Colour coding paperwork can help to differentiate specific procedural systems that are regarded as being particularly prone to error or when the risk of such an error would result in serious consequences. Likewise, controls on machinery should be clearly labelled. Colour and/or different auditory markers can be used to differentiate key components.

- Cognitive overload can be avoided by presenting only the information that is necessary at the time. Long-winded memos and impenetrable text will lead to distractions caused by having to make too great an effort to access the required information. However, supplementary information should be provided at the point where it might be required.

- To avoid unnecessary cognitive effort, wherever possible, forms and paperwork should be designed so that recognition and confirmation of elements replaces the need for recall and extended writing.

- With regards to operating machinery, controls that are similar in design but that perform different functions should not be placed adjacent to each other as this could result in slip-type errors.

- Also with regards to operating machinery, input should result in immediate and unambiguous feedback as to what has been altered so as to allow the operator to undo any changes that have been made.

Mistakes

6.148 Mistakes are errors that result from misunderstanding a process or situation. They are more common than slips and much harder to prevent because the perpetrator will believe that what they are doing is correct until the error occurs. Mistakes are also harder to detect because there are powerful psychological mechanisms that confirm our opinions once adopted, even in the face of contrary evidence.

6.149 A simple but important distinction can be made when analysing mistakes:

- **Errors of commission** are errors that result from action, for example misunderstanding a set of procedures that results in confidential information being sent to the wrong people.

- **Errors of omission** are misunderstandings that involve a lack of activity when an activity is required.

6.150 Mistakes can be further divided according to work-based activities. Put simply:

- **Skill-based errors** are those that occur as a result of inappropriate application of learnt skills or where skills are not fully developed.
- **Rule-based errors** are those that occur as a result of incorrectly or inappropriately followed procedures.
- **Knowledge-based errors** are errors that arise through poor planning or misdiagnosis.

Avoiding mistakes

6.151 Vital to any process of organisational development is information about the state of the organisation. A key to avoiding mistakes, therefore, is to conduct a job/role analysis so that a clear idea is gained as to what job/roles and procedures actually entail. Job/roles described in concrete behavioural terms, ie terms that are observable, controllable and measurable, are far more useful than more abstract definitions. For example, a require-ment 'to be responsible for monitoring risk related behaviour' is not as useful to the organisation or to the employee as a set of clearly defined behaviours that must be carried out. The more detailed and behaviourally based the job analysis is, the more useful it will be for a range of organisational improve-ments related to error reduction and the management of risk.

6.152 However, as the nature of the job/role becomes more complex or senior, and therefore less 'skills' based, so it becomes more difficult to describe 'behavioural' competencies associated with the job/role. For example, a senior manager may possess a wealth of knowledge and understanding that cannot be easily operationalised as a skill to be acquired or displayed.

6.153 A detailed job/role analysis will provide the basis for identifying where errors are likely to occur and the types of errors likely to be made. This can gen-erate modifications in procedures and practices that mitigate against such errors.

6.154 Such a job/role analysis will, in conjunction with a 'skills' audit, also provide the basis for identifying a skills gap to be bridged by appropriate learning and development programmes. The term 'skills' is used in its tradi-tional generic sense; however, as has been discussed, mistakes are not only due to a lack of technical expertise (skills) but also due to incorrectly fol-lowed procedures and a lack of understanding as to the nature of the task. While training can provide technical skills and knowledge of procedures, more senior-level jobs will require the ability to plan and diagnose situations that may not yet have occurred. Such higher-order cognitive processes result from 'education' and 'learning' and will include developing the appropriate attitudes and values that are necessary to carry out the job/role so that mistakes are not made. With regards to the management of risk, different aspects of learning and development will be required at the organisation-wide level, management level and job/role level (see section entitled 'Learning and development').

6.155   It is possible to minimise human error by selecting the right person for the job in the first place. Careful selection procedures based on a job/role analysis to identify the required competencies will help to avoid the wrong person being selected. Selecting the wrong person, particularly in the case of managers, can result in costly mistakes that accumulate over time. From the point of view of human error, inefficiency is largely the result of low-level misdiagnosis and implementation of solutions to a problem. Indirect costs of poor selection processes are associated with a negative influence on morale and job satisfaction.

Latent failures

6.156   Latent failures refer to errors and violations of risk related procedures that have no immediate effect but 'incubate' within the system until triggered. They are very difficult to prevent as it is not possible to foresee every potential error and how they might interact. Reason et al. (1988) distinguish between accidents caused by human factors and those caused by organisational factors. Human factors can be generally thought of as active errors or violations caused by an employee, while organisational factors are considered to be failures of systems and procedures. System failures are likely to be the main cause of latent failures. 'Rather than being the main instigators of accidents, operators tend to be inheritors of 'pathogens' created by poor design, incorrect installation, faulty maintenance, inadequate procedures and management decisions, and the like. The operators' part is usually that of adding the final garnish to a lethal brew that has been long in cooking.' (Reason et al, 1988)

6.157   Catastrophic organisational failures are often caused by latent failures. Turner and Pidgeon (1997) described the incubation period as the time during which there is an accumulation of unnoticed risk factors such as a lack of clear responsibilities for the management of risk, poor communication frameworks, or a poor learning and development programme. Avoiding latent errors is therefore primarily concerned with developing organisational factors so that organisations are resistant to the risk of error.

Deliberate violations

6.158   Deliberate violations of procedures or consciously behaving in ways that ignore the risk of error are not errors. However, violations are likely to increase and compound the consequences of error in organisations where the enforcement of risk management rules has been allowed to drift.

6.159 Reducing deliberate violations of known risk management procedures and encouraging the development of a positive risk culture is essentially what has been discussed in this chapter. In the first place, it is necessary to implement an integrated risk management system that focuses on developing the organisational factors that minimise the risks of loss and accidents, and better control the uncertainties associated with being in business. These organisational factors include: senior management commitment to the risk management process, responsibility frameworks, communication frameworks, job/role design, and learning and development programmes.

6.160 Reducing deliberate violations and developing a positive risk culture also requires an understanding of what causes individuals to commit such actions in the first place, ie the human factors that result in risk related behaviour. The perception of risk, motivation and the nature of human error have been discussed in this context.

REFERENCES

Arnold, J, Cooper C L, and Robertson, I T, *Work Psychology: Understanding Human Behaviour in the Workplace*. London: Financial Times/Pitman Publishing, 1998

Baker, E, Israel, B, and Schurman, S, Role of Control and Support in Occupational Stress: An Integrated Model, *Social, Science and Medicine,* 43, 1145–1159, 1996

Cooper, C L, and Sadri, G, The Impact of Stress Counselling at Work, in Perrewe, P L (ed), *Handbook of Job Stress (Special Issue), Journal of Social Behavior and Personality*, 6(7), 411–423, 1991

Fox, M L, Dwyer, D J, and Ganster, D C, Effects of Stressful Job Demands and Control on Physiological and Attitudinal Outcomes, in a Hospital Setting, *Academy of Management Journal*, 36, 289–318, 1993

Herzberg, F, *Work and the Nature of Man*. Cleveland: World Publishing, 1966

HSE, *The Draft Management Standards for Tackling Work-Related Stress* (www.hse.gov.uk/stress), 2003

HSE, *Real Solutions, Real People: A Manager's Guide To Tackling Work-related Stress*. Sudbury: HSE Books, 2003

Janis, I L, *Victims of Group Think: a Psychological Study of Foreign Policy Decisions and Fiascos*. Boston: Houghton and Mifflin, 1972

Jones, J R, Huxtable, C S, Hodgson, J T, and Price, M J, *Self-reported Work-related Illness in 2001/02: Results from a Household Survey*. Sudbury: HSE Books, 2003

Karasek, R A, and Theorell, T, *Health Work: Stress, Productivity, and the Reconstruction of Working Life*. New York: Basic Books, 1990

Kelly, J E, *Scientific Management, Job Redesign and Work Performance.* London: Academic Press, 1982

Locke, E A, Towards a Theory of Task Motivation and Incentives, *Organisational Behaviour and Human Performance*, 3, 157–189, 1968

Lord Cullen, The Piper Alpha Inquiry, 1990

Lord Cullen, The Cullen Enquiry. The inquiry into the Ladbroke Grove rail accident in 1999, 2001

Morgan, G, *Images of Organisations.* London: Sage Publications, 1997

Naatanen, R, and Summala, H, *Road-User Behaviour and Traffic Accidents.* Amsterdam: North-Holland, 1976

Pickering, T, Job Stress, Control, and Chronic Disease: Moving to the Next Level of Evidence, *Psychosomatic Medicine* 63, 734–736, 2001

Rasmussen, J, What Can Be Learned from Human Error Reports?, in Duncan, K D, Gruneberge, M, and Wallis, D (eds), *Changes in Working Life.* London: Wiley, 1980

Reason, J T, *Human Error.* Cambridge: Cambridge University Press, 1990

Reason, J T, Manstead, A S R, Stradling, S, Baxter, J, Campbell, K, and Huyser, J, *Interim Report on the Investigation of Driver Errors and Violations.* Manchester: Department of Psychology, University of Manchester, 1998

Reynolds, S, Taylor, E, and, Shapiro, D A, Session Impact on Stress Management Training. *Journal of Occupational and Organisational Psychology*, 66, 99–113, 1993

Schein, E H, Organisational Culture, *American Psychologist*, 45(2), 109–119, 1990

Shirom, A, Burnout in Work Organisations, in Cooper, C L, and Robertson, I (eds), *International Review of Industrial and Organisational Psychology.* Chichester, UK: Wiley, 1989

Skinner, B F, *Science and Human Behaviour.* New York: Macmillan, 1938

Turnbull, N, Internal Control – Guidance for Directors on the Combined Code, Report of the Internal Control Working Party, Institute of Charted Accountants in England and Wales, 1999

Turner, B, and Pidgeon, N, *Man-Made Disasters* (2nd edition). London: Butterworth Heinemann, 1999

Wilde, G J S, The Theory of Risk Homeostasis: Implications for Safety and Health, *Risk Analysis*, 2, 209–225, 1982

CHAPTER 7

Getting started – linking OSH and business risk management

INTRODUCTION

7.1 This chapter discusses how OSH professionals can develop a strategy for increasing their contribution to the management of risk within their organisation, which will increase their influence, and contribute significantly towards the performance and competitive advantage of the organisation.

7.2 The development of the strategy will be tackled in two parts:

1   The identification of OSH risks, and how to classify/prioritise them to build an action plan.
2   The development of a strategy and the use of organisation-wide processes to increase influence.

7.3 Earlier in the book we discussed the identification and assessment of risks. In this chapter we will not specify what method of risk assessment or categorisation is used, as organisations may already use a formal system – as part of its organisation-wide risk management system – that can be adopted, so OSH can demonstrate that they are using the same process as other risk management processes. If the organisation does not have an organisation-wide process, then OSH professionals should approach relevant functions within the organisation that are already identified with some aspect of risk management, eg insurance, capital and business continuity, to gain insights and obtain advice about a process that would be appropriate for the OSH function to use.

7.4 The focus of this chapter will be the detailed identification of risks, methods of classification/prioritisation, and showing how OSH professionals can build an effective OSH risk management strategy that can make a significant contribution to the organisation's management of risks and the achievement of its objectives.

7.5 A key theme is that OSH professionals should base their contribution to risk management on the principle that the organisation within which they operate is one large 'system' of interconnecting pieces that together will achieve the organisation's objectives. Our definition, first spelled out in **Chapter 1** is that:

> 'An organisation is a network of relationships between people who come together for a common purpose.'

7.6 Using the key words from the above definition, we can see that the OSH function can support an organisation and its senior management in the following ways:

1   Help to identify and maintain its common purpose, strategy and objectives.
2   Help to manage relationships.
3   Help to maximise opportunities and minimise risks to people in the network and the organisation as a whole.
4   Help to ensure that the people in the organisation have the competencies, knowledge and skills to build effective relationships and successful networks.

7.7 In earlier chapters we have prompted OSH professionals:

- To create initiatives to increase their influence at higher levels of management decision-making.
- To increase their added value, all within a new business and risk-management-focused approach.
- To put aside the organisation's and manager's perception of its contribution and professional image.
- To create a strategy to enhance their professional and personal development.
- To adopt a risk-based approach.

7.8 Also mentioned in earlier chapters is that, if OSH professionals are to 'get on the risk and business agenda' and start to increase their influence and contribution, then they need to identify the 'hooks' that will generate a favourable response from the rest of the organisation. Reducing risk and the exposure to unplanned direct and indirect costs that have financial and other implications for the business will assist the achievement of the organisation's objectives, and increase the contribution and influence of the OSH function.

7.9 It is important to use a structured process for creating a strategy for increasing influence, and in our experience this should be tackled in two stages:

1   Identify the current status, the supporting evidence and the implementation challenges.
2   Create a strategy for increasing influence, and identify processes to assist with the implementation of the strategy.

IDENTIFICATION OF OSH RISKS

7.10   In general terms, there is a need to identify in total where the organisation, the OSH function and OSH professionals (both professionally and personally) are 'at risk', arising from the management of the organisation's OSH risks, and all its related aspects.

7.11   As discussed in earlier chapters, OSH professionals have the opportunity to influence the organisation at the operational, tactical and strategic levels. However, the credibility of OSH is often affected by the internal view of their ability to support operational-level functions and achieve legal compliance. If OSH are unable to get the 'basics right', then any attempt to get involved at the tactical and strategic levels will be resisted.

Identification factors

7.12   We believe that the process of the identification, assessment, prioritisation and treatment of OSH risks should not be different to that used for managing other risks.

7.13   We believe that it is important for OSH professionals to be able to demonstrate to the organisation that a systematic and consistent approach has been used.

7.14   In our experience, the successful management of OSH risks will be affected by four main factors:

- **Factor 1** – OSH strategic, tactical, operational, professional and personal risks.
- **Factor 2** – organisational context within which the organisation is operating, eg the organisation's activities, where they are based, the ownership status.
- **Factor 3** – policies and systems for general risk management.
- **Factor 4** – risks generated by the organisation's business and operational processes, which either have a direct or indirect effect on the management of OSH risks.

7.15   This section will use these four factors to show how risks can be identified. Additionally, at the end of this section we describe a review tool that focuses on these factors. It can be used by OSH professionals to get a 'handle' on the organisation's OSH risk profile, and therefore provide valuable input to the development of a strategy for improving the management of these factors.

Recording the outputs – Risk Register

7.16   It is vital that output from the identification process is recorded. An 'OSH Risk Register' should be used to record all OSH risks that are identified and any potential risk treatments. Any measurement and performance data that is available, eg accident/incident figures, absence rates, property damage costs should also be included to assist with the assessment of the risk in later steps. See **Chapter 2** for an indication of typical data and reporting methods.

7.17   In particular the Risk Register should include all the risks, and potential risk treatments arising from the consideration of the four factors described below, plus a review of the OSH business processes outlined in **Chapter 5**.

7.18   The Risk Register should provide an overview of the status of the organisation's management of its OSH risks, and start the process of identifying how the management system can be improved.

FACTOR 1: IDENTIFY RISKS – ORGANISATIONAL CONTEXT

7.19   The best place to start is by undertaking an exercise to identify organisational factors and the organisational context within which the organisation and the OSH function have to operate. A list of organisational factors, taken from **Chapter 1**, and used as the structure for this factor, is set out below:

- Vision statement.
- Senior management commitment.
- Core values.
- Leadership style.
- Organisational culture.
- Hidden belief systems.
- Psychological contract.
- Responsibility framework.
- Occupational stress.
- Job/role design.
- Business strategy and goal-setting process.
- Performance management system.
- Compensation and rewards policies.
- Communication.
- Learning and development.

7.20   We have provided some explanation of each organisational factor and its relevance to risk identification, and under each factor listed a series of questions. We recommend that the whole OSH function be involved in creating a consensus about the answers, and only when a consensus is agreed should the risk be entered into a Risk Register. As a starting point, if the answer to any of the questions is 'no', then you have created an item for your Risk Register.

- **Vision statement** – a statement of the organisation's vision for its future (sometimes called a 'mission statement') that underpins everything it says and does. Many organisations create vision statements,

but develop the words as if the exercise was part of their public relations or general marketing/sales campaign. In the absence of a clear and demonstrable vision, the organisation lacks a clear statement around which many other things need to revolve. For example, if an organisation says that its vision is 'To deliver outstanding customer value, provide exceptional customer service, create excellent returns for its shareholders, and make itself the employer of choice', then all of its value statements, strategy, goals and objectives must flow from that vision. However, many organisations will publish such a vision, and not engage people in establishing the values and behaviours that will ensure the vision can be achieved. Or they will publish a vision statement, and the associated values, but not link them to the organisation's strategy, goals and objectives and leadership competencies.
Questions:

1   Does your organisation have a vision (mission) statement?
2   Is it clear, demonstrable and well-communicated?
3   Does it underpin everything that the organisation says and does?
4   Is it linked to the organisation's strategy, goals and objectives?
5   Are there clear, demonstrable and well-communicated value statements that are derived from the vision statement?

- **Senior management commitment** – the extent to which management are committed to the management of risk and the management of an organisation's OSH risks.
  Questions:

  1   Do senior management actively and publicly support the management of risks, including OSH risks?
  2   Do senior management take part in OSH risk management project teams?
  3   Do senior management discuss risk management, internal control processes and OSH risks at main management meetings?
  4   Do senior management involve themselves in reviews to establish the current status and past performance of the organisation's risk management, internal control processes and OSH risks?
  5   Do senior management performance measurements include those related to the management of risk, internal control and OSH?

- **Core values** – as we have discussed above, core values need to derive from the organisation's vision statement. Core values describe what particular aspects the organisation focuses on, what core values it sets for the organisation and its employees.
  Questions:

  1   Does your organisation have core values?
  2   Are they clear, demonstrable and well-communicated?
  3   Do people accept the core values?
  4   Do they create the attitudes/behaviours that the organisation requires?

    5   Are definitions of leadership behaviours, which demonstrate the values, included in the personal development planning process?

    6   Does the organisation's strategy, goals and objectives fit with the core values?

- **Leadership style** – the energy and inspirational qualities of an organisation's leaders are a major factor in creating an effective organisation.

Questions:

1   Do they create and live by the organisation's vision and core values?

2   Do they run the organisation on strong values and principles, especially related to employees, customers and suppliers?

3   Do they genuinely care for their people?

4   Do they involve their people and give them freedom?

5   Do they show appreciation and ensure that work is fun?

6   Do they show real trust?

7   Do they listen and act on what they hear?

- **Organisational culture** – what value does the organisation place on the effective management of risk and in particular OSH risks?

Questions:

1   Does your organisation see the management of risk as a valuable process that can protect all stakeholders?

2   Does the organisation accept that considered risk-taking is part of creating a vibrant and dynamic organisation?

3   Is the taking of risks seen as a core value of the organisation?

4   Will people who take intelligent risks be 'rewarded', rather than blamed, when planned outcomes are not achieved?

5   Does your organisation give a clear lead on the management of its OSH risks?

6   Are OSH professionals able to proactively influence the organisation's strategy, goals and objectives?

7   Are OSH professionals part of the organisation's business and commercial processes, including the management of risk?

- **Hidden belief systems** – what hidden belief systems are actually in place, and what influences attitudes and behaviour when no formal 'rules' are available?

Questions:

1   Does your organisation have hidden belief systems that are different to the organisation's vision and core values?

2   Do the organisation's vision and values take precedence in influencing attitudes/behaviours?

3   Do your senior management recognise that the organisation has a hidden belief system?

4   Could your senior management describe in writing the content of the organisation's hidden belief system?

5   If the current hidden beliefs are contrary to the organisation's vision and values, would your senior management accept the need to create a new set of beliefs that could support the organisations vision and values?

6   What information do you have on levels of trust in the organisation?

- **Psychological contract** – the psychological contract is an unwritten set of expectations between everyone in an organisation and, unlike the written contract, is continually changing. Of particular significance to the management of OSH risks is the erosion of trust between employers and employed.

Questions:

1   Does the organisation recognise the existence of a psychological contract?

2   Would management be able to describe, in writing, their interpretation of the contract and its value to the organisation?

3   What information do you have/need on the quality of relationships and trust in the organisation?

4   Has the status of the contract remained static in recent times?

5   Have the changes been for 'the better', ie have they improved confidence and trust between employees and management?

- **Responsibility framework** – to what extent are responsibility, authority and the allocation of resources delegated to the correct level at which action is most effective?

Questions:

1   Is responsibility and authority clearly defined and communicated?

2   Are decisions taken at the most effective level?

3   Are managers effective at delegating responsibility and authority?

4   Does budgetary authority match general management authority?

- **Occupational stress** – occupational stress has many negative consequences for organisations, and therefore it is a strategic risk that can be identified, quantified and managed. Numerous studies have shown that workplace stress adversely affects work performance, morale and commitment to the organisation.

Questions:

1   Has the organisation used absence, health monitoring and other data to identify the existence of occupational stress?

2   Does the data show how and why occupational stress is affecting the organisation?

3   Has the organisation undertaken any audits and/or employee surveys to identify the existence of occupational stressors?

4   Have the results of any audit/surveys been built into an effective action plan?

5   Is there a current system for identifying, at an individual employee level, the presence of occupational stress?

6   Have managers been trained to recognise and manage occupational stress?

- **Job/role design** – the ability of an organisation to create a structure, management style and work environment that encourages the best aspects of team working is a vital component in successfully managing any activity within the business, especially risk management.
  Questions:
  1   Has the organisation set clear guidance for the design of jobs/roles and team working?
  2   Is the management structure aligned to the jobs and roles that the organisation has created?
  3   Does the management structure enable people to fulfil roles that will enhance the achievement of organisational objectives?
  4   Is active communication outside of formal management structures and team working naturally encouraged?
  5   Are active communications and effective team working rewarded over individual performance?

- **Business strategy and goal-setting process** – to what extent are these elements integrated and are goals and objectives formally set, cascaded, monitored and amended to take account of changing circumstances?
  Questions:
  1   Are the organisation's vision, values, strategy, goals and objectives all interlinked?
  2   Is there an effective formal process for communicating about and cascading the organisation's strategy, goals and objectives so plans can be developed at the most effective level?
  3   Are managers allowed latitude in developing plans that meet their goals and objectives, rather than being imposed from higher levels?
  4   Are the plans used to create performance indicators that are used during performance reviews, and related to reward and recognition processes?

- **Performance management system** – to what extent are performance goals related to OSH risk management agreed, measured and managed to ensure active action-oriented activity?
  Questions:
  1   Does the organisation require that regular performance review and development discussions occur for each employee?
  2   Do managers regularly discuss performance and development with their people?
  3   Have participants received training in the system?
  4   Are managers rewarded for being good people developers? Are penalties imposed on managers if they do not undertake regular performance reviews?

5 Is the process used a genuine two-way dialogue rather than an imposed result determined by the manager?

6 Is anyone other than the employees' manager required to 'sign-off' the results?

7 Is the employee given an opportunity to 'appeal' against the results of performance assessment?

8 Are performance measures determined and agreed with the employee?

- **Compensation and reward policies** – to what extent are compensation and reward policies aligned to the business, linked to business goals and objectives, including OSH risk management; and are those behaviours that are consistent with the organisation's core values, cultural framework and hidden belief systems rewarded?

Questions:

1 Are your compensation and reward policies linked to the organisation's vision, values, strategy, goals and objectives?

2 Are they designed to produce specific attitudes/behaviours that reward the employee for the actions required?

3 Have those attitudes/behaviours been created?

4 Have they achieved the desired result?

5 Do the policies provide an equitable distribution of the benefits generated by the policy?

6 Are individual performance measures related to reward and recognition processes?

- **Communications** – ensuring that communication systems are effective is a challenge in most organisations but the issue becomes ever more relevant when related to the management of risk, and in particular OSH risks. Effective communication is the foundation on which risk management systems are based.

Questions:

1 Is there a formal communications strategy?

2 Are there different methods used to distribute only essential information to those who need to act upon it?

3 Are any 'mass' communication systems focused on 'face-to-face' and active involvement?

4 Are communications methods 'two-way', and are there effective feedback arrangements?

5 Are the matters raised during communication actively recorded, and is timely feedback provided?

6 Are regular employee surveys used to gain understanding of employees' issues and concerns and alignment with the vision, values and strategy?

7 Is the information gained from these used to engage employees at work-group level in discussions on how to improve areas of weakness?

- **Learning and development** – to what extent are the organisational and individual needs for learning and development considered as part of the means by which the organisation will achieve its objectives, and a key method used to enhance the contribution and productivity of the individual employee and teams, and their effective contribution to OSH risk management?

  Questions:

  1 Are there formal systems to identify organisational core learning needs, which include OSH requirements?

  2 Do individual performance reviews identify personal development needs?

  3 Do managers and employees accept mutual accountability for delivering the development plan?

  4 Are these compared with and amalgamated with the organisational core requirements to produce a complete organisational perspective and priorities?

  5 Based on those priorities, is an annual learning and development plan created and adequate resources allocated?

  6 Are the results of learning and development expenditures formally evaluated to assess their contribution to the organisation and individual?

7.21 Additional organisational factors that may be considered are:

- Organisational activity.
- Locations of sites/people.
- History/background of organisation.
- Ownership.
- Particular political and external influences.
- Exit plans (where appropriate).
- Organisational structure/style.
- Strategies/goals/objectives.
- Commercial status.
- Internal control systems.
- Management control systems.

7.22 These factors will themselves vary considerably between and within organisations, depending on a whole variety of factors. Consequently, the impact of the organisational context on the management of its people risks will also vary. OSH professionals need to identify the factors that are relevant to their organisation, evaluate the impact, and record any additional risks in the Risk Register. As mentioned earlier, at the end of this section we will give details of a tool that can be used to focus in on the organisational context factors that are particularly relevant to the management of people risks within an organisation.

FACTOR 2: IDENTIFY RISKS – RISK MANAGEMENT PROCESSES

7.23 The policy, strategy and processes that already exist within an organisation for general risk management, internal control and corporate governance will have a greater or lesser impact on the management of OSH risks, depending on the extent, coverage and sophistication of existing systems. There is a clear need for OSH professionals to adopt a structured and systematic process that takes account of existing systems, especially a risk management vocabulary, methods of risk assessment, classification of risks, and project management. The three main areas that OSH professionals should consider are any existing arrangements for:

1   Risk management strategy and organisational framework.
2   Internal controls and processes.
3   Project management methods.

7.24 OSH professionals need to discuss with existing 'risk management' functions, eg insurance, business continuity, to identify existing arrangements and create and dovetail an approach that is consistent and does not produce conflict, misunderstanding or confusion.

7.25 The two risk management standards described in **Chapter 1** provide some excellent guidance for the development of a framework for an organisational process for managing risk. Additionally, we have explained about organisational systems for business risk management in earlier chapters.

FACTOR 3: IDENTIFY RISKS – OSH RISKS

7.26 The next step is to use a systematic process to identify OSH risks that are:

●   Created by an organisation's management of its OSH risks.
●   Created by the activities of the OSH function.
●   Potentially reduced/improved by the input of the OSH function.

7.27 We see these risks falling into two main categories:

1   OSH risks at three levels – strategic, tactical and operational.
2   OSH professional and personal risks.

OSH risks

7.28 The first step to identify operational risks is to look at any previous audit reports that have been undertaken on the OSH management system, the OSH function, policies, procedures and record systems.

Audit

7.29   If the organisation's management of its OSH risks has not been audited, then serious consideration should be given to undertaking an audit using an external auditor, either external to the function or external to the organisation. An audit will provide a very useful insight into the current performance level, but need not study every activity or process in detail. In our experience, the most effective audits are those that combine 'horizontal' and 'vertical' processes. By 'horizontal' we mean a study horizontally across the whole function, at a certain level of activity, eg operational, tactical or strategic. By 'vertical' we mean a study of a particular activity, eg hazard control from 'top to bottom' or 'beginning to end'. We recommend that you use a 'horizontal' approach at the strategic and tactical levels, and a 'vertical' approach for a study of the main operational activities. Remember that, during the audit, especially the 'vertical' elements, that you actually look for any evidence rather than assuming it 'will be here somewhere'! The evidence of systems performance on 'paper' should be supported and validated by a review of physical evidence, including the compliance to standards, eg housekeeping and the compliance of employees to personal protective equipment requirements. These two indicators can identify whether the organisation's systems for managing OSH risks have been implemented effectively and subsequently monitored.

7.30   In addition, we recommend that every opportunity be taken to identify and capture any performance data that can be used to assist with the risk assessment process. The audit results and data can be used to create entries for the OSH Risk Register.

7.31   If an audit has been undertaken, the OSH function needs to ensure that the results and recommendations have been fully considered and that an action plan is in place to ensure the recommendations are being implemented. If an action plan does not exist and some recommendations have not been implemented, then those items should be entered in the OSH Risk Register. In addition, any performance data or analysis that can be used for the risk assessment step should be added into the Risk Register.

Input, activity, output approach

7.32   In addition to thinking about the risks that can be created by a particular activity, it is equally vital to identify the risks that are initially caused by factors that affect a particular activity – called 'input risks'. There is also a need to consider the 'knock-on' effect of the actual activity or organisational aspect – called 'output risks'. In addition, when thinking about the organisation as a complete system it is clear that 'input risks' to a particular activity or aspect are generally 'output risks' from another part of the system.

7.33 Figure 7.1 demonstrates the relationship between input risks, activity risks and output risks.

FIGURE 7.1

7.34 Too often, the area for risk identification is drawn far too tightly, and misses the realities of the risks being created by a business activity. In our experience, the identification process is frequently restricted to the 'activity' risks, and the risks created by the inputs and those transferred to the outputs are usually not even considered. In addition, the 'Activity' itself is often assessed at a superficial level, based on the assumptions of what is taking place and the inherent risks rather than on detailed observation of what actually happens.

7.35 The reality is often very different. This case study shows the risks that arise when a particular activity (in this case the installation/commissioning of a piece of equipment) does not proceed to the plan agreed with the sales representative, and the service department have to 'pick up the pieces'!

Case study

7.36 An organisation sells and services a production machine that can be used for many different purposes.

INPUTS

7.37 Following a call to customer services, a sales representative visits the potential customer to discuss their needs. The sales representative discusses various options and views the plans of the intended installation site, but does not actually visit the location to identify the risks. When asked whether the equipment can be installed in the location shown on the plan, the sales representative confirms that the location is acceptable, as it is away from other machinery. The sales representative does not, however, volunteer information about installation/commissioning requirements or servicing costs, as this will potentially put the sale at risk. In particular, no information is volunteered about the equipment needing to be connected to certain services, including extraction, and the cost of servicing.

ACTIVITY

7.38   The customer orders the equipment and it is delivered to site. The installation/commissioning team arrives on site to discover that to connect the equipment to services within its design parameters and to maintain its operational efficiency will require a different location, or the installation of new service connection points closer to the intended location – with an obvious delay. The customer is not pleased with the possibility of extra cost (believing that the purchase price covered the whole costs), or further delay, so agrees to a new location close to existing service connections, which are in the same area as existing production machinery that cannot be shut down at short notice, plus there is a lot of pressure to maximise production. Although to continue with the installation/commissioning would place the team at risk, as they are on a bonus scheme linked to the number of completed installations, they decide to proceed making sure that each 'watches out for the other' to prevent any risks arising. No risk assessment or safe system of work is developed and agreed with the customer. The final testing is carried out remotely from the new equipment, the machine is fully installed and commissioned, and the installation team leave the site.

OUTPUTS

7.39   After the warranty has expired, the customer services department receives a call from the customer about an intermittent fault with the new equipment. The customer has previously declined the offer of a service contract. The customer asks for a price for the service visit, and comments that they are getting quotes from other companies. As the customer service department is on an incentive scheme to maximise chargeable service work, the customer service representative does not want to lose the 'sale', so gives the customer a price based on a single engineer visiting the site and completing the repair within two hours, without asking any questions about the equipment/location, etc. Standard question sets are not used. The customer agrees to the price and a single service engineer is despatched. The service engineers are also on a bonus scheme, but this relates to minimising the time spent on service visits, relative to chargeable hours, so additional service visits can be completed.

7.40   On arrival, the service engineer discovers that the installation position cannot be reached without the other machinery being shut down, which the customer will not agree to because of production delays. However, as the fault is only intermittent, the service engineer decides to start the service call believing that the fault can be rectified remotely. That does not work so, to prevent a further service call, the engineer starts to work on the new piece of equipment. During the servicing activity, the service engineer becomes trapped by the other machinery, and is seriously injured.

Case study risks

7.41   The risks under each category can be summarised as follows.

7.42   Input risks:

- Sales representative's desire to 'make the sale' without discussing all the requirements of the machine.
- Bonus scheme focused on personal sales volume, with no element related to sales process.
- No process/checklist to ensure discussion of installation, commissioning and servicing.
- No details in contract about installation/commissioning/servicing requirements or the potential need for a safe system of work and shut down of existing machinery during these activities.
- No prior site survey to identify risks and requirements for installation/commissioning activity.
- No prior discussion/agreement with customer about actual product requirements.

7.43   Activity risks:

- Agreeing to locate machine in unsafe area.
- Proceeding with activity without safe system of work and shutdown of existing equipment.
- No record made of machine location.
- No record made of issues arising from the activity.
- No information passed on to customer service or service department about the installation/commissioning challenges in relation to OSH.
- Bonus scheme that focused on 'productivity' only.

7.44   Output risks:

- Customer service representative on bonus scheme to maximise 'sales', with no process element.
- Customer service does not have a process to ensure that all relevant information is obtained before a quote is provided.
- Very limited information is passed to service engineer.
- Service engineer is on an incentive scheme to maximise service calls.
- Service engineer does not undertake risk assessment before proceeding.
- Service engineer does not make the customer aware of the risk with the servicing, based on the machine's location.

7.45   We recommend that this case study be used to prompt the creation of a list of 'input-activity-output' exposures that are related to the OSH management system. Each potential 'input-activity-output' exposure should then be analysed to identify the risks. The risks should be added to the Risk Register, together with any data for the risk assessment process, and details of potential risk treatments.

OSH professional and personal risks

7.46   In earlier chapters we have discussed the way that OSH professionals can help an organisation to achieve its objectives. In this section we have listed some professional and personal risks that are important in understanding how the contribution of the OSH function is restricted, and what potential risk treatments are available. The list is not intended to be exhaustive, and should be used by OSH professionals to prompt the generation of a list that is relevant to a particular organisational context.

OSH professional risks

7.47   Professional factors that can create risks for OSH include:

- What is the organisation's opinion of the contribution made by OSH to the achievement of the organisation's objectives?
- What is the organisation's opinion of the OSH function, and what areas does the organisation think OSH does well or could be improved?
- What is the organisation's perception of the ability of the OSH function to increase its influence, and in what areas?
- Being too closely aligned to the management process, therefore not able to provide impartial advice or take a balanced view.
- Being seen by the organisation and 'general' management as a legal compliance risk-averse function, thereby not recognising or asking for a more effective contribution.
- Focusing on operational level functions, thereby restricting their ability to make a more effective contribution.
- Being seen as a 'blocker' to progress and a guardian of a risk-averse legal compliance activity.
- Seeing themselves in the role of legal compliance enforcer.
- Not 'speaking' a business language, and not making a contribution to business and commercial strategy, goals and objectives.
- Focusing on the cost of intervention and not considering the cost of inaction.
- Focusing only on the minimisation of risks/costs and not including a focus on the maximisation of opportunities.

- Not using available data to identify risks and to influence methods of cost control.

- Not maintaining an active continuing professional development (CPD) process for those working in the OSH function, which covers both professional and personal development.

7.48   Add any identified risks to the Risk Register, along with any 'evidence' or data, plus consider potential risk treatments.

7.49   In addition, the OSH function should create a professional development plan that covers the whole function and is linked to the current roles that members of the team undertake. Each team member should identify where professional development opportunities exist and where their relationship with, and support for, parts of the organisation or the management of particular support activities can be enhanced. In particular, input should be sought from other managers within the organisation. The obstacles that are preventing such developments being implemented currently should also be identified. These individual development opportunities and obstacles should be amalgamated to create a schedule for the whole function, that can then be used to identify priorities and key actions.

OSH personal risks

7.50   OSH professionals are no different to any other job-holder, in that their contribution to an organisation can be restricted by the 'constraints' of their current role or the historical perception of the abilities of previous job-holders. However, to ensure that the job or the historical perception does not constrain a particular job-holder, the application of a different blend of personal competencies, skills and experience must be brought to bear so the individual can 'rise above' the expected contribution. The personal risks that can affect an organisation's management of OSH risks can be tackled by understanding what competencies, skills and experience are required and what development opportunities are available.

7.51   OSH professionals need to develop the key skills of facilitation and relationship building to ensure that the organisation maintains an open culture and challenging corporate 'mind' and is prepared to operate as a matrix of opportunity 'teams', rather than each activity within the organisation protecting its own responsibility areas and not thinking about or contributing to the overall strategic objectives. In addition, they need to develop and bring to bear the skills of generic influencing, facilitation, blue-sky thinking and non-silo-constrained skills to support the development of business-focused and commercially relevant OSH risk management systems. However, the OSH professional will often need allies within the organisation to start the process of getting on the risk and business agenda.

7.52   The Code of Professional Conduct published by The Chartered Institution of Occupational Safety and Health (IOSH), Europe's leading body for OSH professionals, states the following.

7.53   Code Point 1:

'Members of IOSH, wherever employed, owe a primary loyalty to the workpeople and the community they serve and the environment they affect. Their practice should be performed according to the highest standards and ethical principles, maintaining respect for human dignity. OSH practitioners shall seek to ensure professional independence in the execution of their functions.'

7.54   The guidance on this point is:

'The term professional independence relates to the function of OSH practitioners within the organisation in which they practice. As a professional they should be able to exercise their OSH function according to their independent judgement.'

7.55   Code Point 4:

'Members shall take all reasonable steps to obtain, maintain and develop their professional competency by attention to new developments in occupational safety and health and shall encourage others working under their supervision to do so.'

7.56   The guidance on this point is:

'Competency is defined in the approved code of practice to the Management of OSH at Work Regulations 1999 [MHSWR 1999] as:

"The possession of sufficient knowledge, experience and skill to enable the person to know what he or she is doing and to be able to carry out a task in the way in which a person competent in the activity would expect it to be done and to have an appreciation of one's own limitations."

Professional competency goes beyond this and is gained by a combination of qualifications and practical experience, at an appropriate level sometimes supplemented by membership of other specialist bodies.'

7.57   On its website (www.iosh.co.uk), the IOSH states that:

'Flexibility and adaptability are the hallmarks of an ambitious (safety and health) practitioner's career. Working within occupational safety and health offers a responsible, professional job with opportunities for talents at all levels. A practitioner may manage people, materials and finance, dealing with practical matters that really count in the working world. You need to be an innovative problem solver and have the right qualities to cope in a crisis. Attitudes and behaviour are factors which may be involved in the cause of accidents. So as a safety and health practitioner you have to discover – and influence – the way people think and behave. That means you must be able to communicate at all levels, as well as demonstrate your technical skills.'

7.58 They go on to talk about their continuing professional development scheme (CPD):

'Participation in CPD is an acknowledgement of the professional's responsibility to industry and the public. Evidence of systematic development is particularly relevant to the occupational safety and health practitioner. In a profession characterised by rapid changes it is vital to be able to demonstrate development of relevant knowledge, skills and areas of competence beyond initial qualifications. CPD provides a structured approach to developing new knowledge to fulfil new responsibilities. Employers are legally required to have access to competent safety and health advice, and evidence of CPD indicates a certain level of competence, as does membership of the Institution of Occupational Safety and Health's (IOSH's) Register of Safety Practitioners (RSP), for which CPD is mandatory. RSPs have an ethical duty to maintain, update and enhance knowledge and competence to enable them to practice effectively in a highly demanding area of work.'

7.59 The IOSH scheme is open to all members of the Institution on an equal basis. It runs in two-year cycles, with participants required to obtain 20 CPD points in each cycle. The points system is weighted to reflect the degree to which activities enhance professional development. Points apply equally to management and technical skills and competencies. The scheme gives each participant the opportunity to plan a pathway towards increased effectiveness and competence, and to monitor progress made. The pathway can be adjusted to take account of new developments.

7.60 There are six categories of CPD activities:

1   Continuing education.
2   Attendance of conferences, seminars and workshops.
3   Presentation of papers or contribution to professional publications.
4   Technical service.
5   Self-development.
6   Preparation of training material; development/implementation of OSH strategies.

7.61 Additionally, the scheme requires details of the learning and development that a participant in the scheme has achieved via the particular activity. The scheme is certainly not one of just collecting points; it is more about what has been learned from the process and how that learning has been applied in their work environment.

7.62 Even if an OSH professional is not a member of IOSH, the above list of CPD activities is an excellent way to consider what personal development each member of the organisation's OSH team requires, equipping them to enhance their added value to the organisation. We therefore recommend that the above list be used to develop a personal risk assessment that can be fed into a personal development plan and CPD objectives, and listed in the Risk Register.

7.63 In this way, the organisation will see that the OSH function is considering all factors to increase its contribution, including personal development. A programme of CPD can also include professional and skills courses that have relevance for the individual and the professional developments needs of the OSH function. Of particular relevance would be more sessions on more general management topics, eg business risk management, people management, operating and financial review reports, corporate governance, corporate social responsibility.

FACTOR 4: IDENTIFY RISKS – OTHER RELATED RISKS

7.64 This area of risk identification needs to consider all the organisation's business and operational risk areas that have a direct or indirect impact on the management of OSH risks and the service that the OSH function provides to the rest of the organisation. If this factor is not included, then a significant number of risks could remain undetected (until something goes wrong!), plus managers in those areas can take the view that the OSH function is only really interested in those aspects that relate to the core function of OSH, and not those that relate to the general operations of the organisation.

7.65 Many of the risks should have been identified during earlier reviews, especially when using the 'input-activity-output' method. However, to ensure that no risk has been missed, we recommend that you list all the organisation's business and operational risk areas and activities and identify the links to the management of OSH risks, and the activities of the OSH function. Having created the list, you will then be able to use the 'input-activity-output' method in new areas of investigation to identify additional risks.

7.66 The table shows an example list of business and operational risk areas.

Business strategy and planning	Sales and service	Facility and plant management
Reputation and legal	Distribution and transport	OSH, quality and environment
Capital and insurance	Procurement and supply chain	IT, fraud and security
Marketing and development	Operations and production	Business continuity

7.67 Add any identified risks to the Risk Register, along with any 'evidence' or data, plus consider potential risk treatments.

Risk Register

7.68 This example of a Risk Register is designed to prompt the identification of OSH risks. It is not an exhaustive list and should only be used as a base from which to start the identification process. This example Risk Register can be found on pages 244–249.

RISK-PROFILING TOOL

7.69 We have included details of a unique risk-profiling tool that we have developed especially for OSH professionals, to assist them in identifying OSH risks.

7.70 The tool is called 'OSH Risk Management' and is the copyright of RiskFrisk®.

7.71 This unique tool is used to undertake a top-level risk-profiling review. It focuses on the way in which an organisation's risks and the interrelationship between them can have an impact on the management of its OSH risks and the OSH function. It reviews four major factors:

- **Factor 1:** OSH risks, OSH professional and personal risks.
- **Factor 2:** organisational context.
- **Factor 3:** general risk management policies, strategies and processes.
- **Factor 4:** business and operational process risks.

7.72 The methodology adopts a holistic approach and includes the 'input-activity-output' method.

7.73 Benefits to OSH professionals are:

- Knowing the key areas where their organisation is at risk.
- Being able to use the outputs to support informed decision-making about priorities for improving the management of OSH risks.
- Understanding the options for managing OSH risk exposures and costs.
- Understanding OSH strategic, tactical, operational, professional and personal risks.
- Understanding how risks in other areas of the organisation relate to and influence the management of OSH risks.
- Understanding how OSH can influence the management of related risks.

7.74 It also provides OSH professionals with:

- Valuable insights into key organisational processes.

Risk area	Potential risk exposures and assessment data	Potential risk treatments
Vision and values and organisational 'culture'	OSH is not a major organisational focus, despite statements to the contrary	Include effective OSH management as a core value
Corporate governance	Managers/employees acting in an inappropriate manner and/or not following procedures. OSH management is not 'on-the-agenda'	Ensure that adequate policies and procedures are in place that include OSH
Strategic planning	Reactive management decision-making. No process for reviewing medium/long-term organisational strategy, goals and objectives	Implement allocation of authority and responsibility, including OSH management
Organisation design and development	Design does not fit the current requirements, and authority and responsibility allocation, especially for OSH is not clear	Establish clear lines and areas of authority and responsibility, especially for OSH Communicate authority and responsibility allocation to all likely to be affected
Employee satisfaction and commitment Communication processes	Employees' views are not considered. Employees are potentially de-motivated, with a reduction in productivity and an increase in absence, accidents and turnover	Implement a two-way communication process that enables employees' views to be proactively obtained, especially on OSH Only use proportionate communications methods to 'target' required audience, ie only send them what they need to know. Do not send everything to everyone

Building trust, effective working relationships and team working	Employees have no trust in the ability of the management to manage	Instil in senior management a culture where they 'walk the talk'
	Individual employees are unsure of their authority and responsibility limits, which stifles creative thinking and decision-making	Ensure that the allocation of authority and responsibility – especially for OSH – is clear and unambiguous, and is well communicated to all concerned
Performance assessment	System of performance assessment focuses on financial/commercial targets, with no reference to OSH	Ensure that system includes OSH
Recruitment and selection process	No inclusion of OSH as part of job requirements, and no inclusion of these elements in recruitment process	Review existing process to ensure OSH and risk management is a core element
Reward and recognition	Reward and recognition systems make no reference to OSH	Ensure that the system is designed to produce the attitudes/behaviours desired, including OSH
Core skills training	No identification of core skills requirements	Ensure core skills process includes OSH requirements at all levels
	No register of internal core skill capabilities	
Harassment, bullying or violence in the workplace	Managers/employees acting in an inappropriate manner	Ensure adequate policies, procedures and training are developed following a risk assessment
	No recognition of the implications for OSH and the need for risk assessment	

Risk area	Potential risk exposures and assessment data	Potential risk treatments
Stress	Employee health and performance is adversely affected, with a reduction in 'productivity', and increase in turnover, absence, etc	Create management systems to identify individual and organisational stressors, to facilitate action planning and case handling
	No management awareness or system to identify and resolve causes of stress	Ensure adequate policies and procedures
	Inappropriate policies and procedures are in place	Ensure that organisational values, policies and approach are well communicated to all, via initial induction/training and ongoing processes
	Managers/employees acting in an inappropriate manner and/or not following the procedures	Ensure awareness training and guidance in case handling is provided to managers
	Employment tribunal cases/costs	
Absence management	Limited system, with no link to accident reporting, especially work-related driving accidents that may also cause personal injury	Ensure adequate system for data capture and retrieval. Link to accident management system
		Analyse data and provide proactive advice to managers
OSH legislation	Failure to identify legal changes	Use an external notification system to identify changes
	Organisation unaware of the need for or	

	not prepared to implement compliant policies and procedures	Use all 'arguments' to demonstrate the need to implement and ensure compliance
	H&S failure to advocate adequate policies and procedures	Ensure full understanding of the requirements via a process of CPD for the OSH function
	Failure to communicate and train those responsible for implementation and compliance	Ensure 'new' requirements are built into manager/employee communication and training processes
H&S business related processes (see **Chapter 5**)	Existing processes do not cover all areas and/or are not related to needs of organisation	Undertake a systematic survey to identify current status and what is required, with particular reference to risk assessment
	OSH function does not communicate the actual needs to managers/employees	Create policies and procedures that will support the organisation to achieve its strategy, goals and objectives, and achieve compliance
	Organisation uses wrong focus for compliance, with the OSH function managing the process, not line managers	Ensure all legally required documentation is created, implemented and updated
	Managers/employees acting without formal guidance and in an unlawful or inappropriate manner	
	Accident costs and other losses	

Risk area	Potential risk exposures and assessment data	Potential risk treatments
OSH functional operations	Incomplete documentation and inadequate training create inconsistent and inaccurate advice	Ensure that all internal policies and procedures are fully documented and up to date
	Issues of interpretation and application are resolved 'locally' and not passed to the 'centre' for review	Create and implement training and development programmes so all H&S employees fully understand the content and application of H&S policies and procedures
	Lack of control creates inaccurate record systems and incomplete data from which decisions can be made	Ensure a record of issues identified, so that amendments can be considered and acted upon
	No formal process of professional/ personal development	Ensure that a system of CPD is implemented for all members of the OSH function
OSH practice	Managers/employees acting in an unsafe manner	Ensure that organisational values, policies and approach are well communicated to all, via initial induction/training and ongoing processes
	Reduction in productivity	
	Reduction in morale due to accidents and inadequate organisational focus	

Occupational health	New employees may have pre-existing conditions that can be exacerbated by their employment Existing employees can be adversely affected by work activities	Use pre-employment medicals, ongoing health monitoring and absence data to identify exposures
Accidents	Accidents at work and 'work-related' driving accidents that cause personal injury are reported as sickness absence, and not linked to actual 'accident' Prosecution cases and costs	Ensure managers know the distinction between the various types of absence and how each has to be handled and reported
Disciplinary and grievance procedures	OSH requirements are not linked to grievance and disciplinary procedures Inconsistent treatment of non-compliance to OSH policies and procedures, etc	Ensure that grievance and discipline procedures include OSH, in particular the reporting of concerns about OSH arrangements under the grievance procedure and the discipline of employees under the discipline procedure for non-compliance to OSH policies, etc

- A view of the organisation as a complete system.
- An opportunity to feed the results into the organisation's strategy and plans.
- An awareness of current silo thinking.
- Information to prompt a review of insurance and other costs.
- An ability to reassure shareholders.
- Knowledge to help support organisational development.

7.75   The business-focused and commercially-relevant outputs are specifically designed to enable OSH professionals to demonstrate where improvements to OSH risk management can be made. Additionally, the output can be used by OSH professionals to demonstrate an understanding of the interrelationships between risk areas, and an understanding of the organisation of a complete system. This gives OSH professionals an excellent opportunity to increase their influence and professional standing within the organisation by demonstrating that they are aware of the broader implications and can more easily provide advice about the opportunities and risks facing the organisation.

7.76   OSH professionals can use the tool to:

- Review their organisation as a total system.
- Demonstrate business and commercial understanding and relevance.
- Demonstrate an integrated approach, and not just a focus on compliance.
- Identify the impact of existing organisational risk factors.
- Identify OSH strategic, tactical, operational, professional and personal risks.
- Identify management system status and required improvements.
- Become aware of (a) actual activity risks, (b) input risks (risks that affect the actual process), and (c) output risks (risks that are created by the actual process and become an input risk for another process).
- Improve the awareness of the specific risk area and its interrelationships, and how changes in one process need to consider the implications for other processes.
- Obtain vital information to enable them to propose business-relevant solutions that are focused on the cost-benefit analysis of an improvement rather than just concentrating on the intervention costs.
- They can initially use the tool to set the scene and establish a baseline position statement and profile.

Ongoing usage

7.77   The use of the tool is not a single event. OSH professionals can apply the process to future changes that are being considered by their organisation, that require a fresh look at risk management and internal control systems and especially the implications for OSH risks. In this way OSH professionals can contribute to the 'risk and business agenda' on an ongoing basis. This inevitably results in OSH professionals demonstrating their business and risk credentials, and their added value to the organisation.

STRATEGY FOR INCREASING THE OSH CONTRIBUTION

7.78   In this section we will:

- Consider the development of a strategy to increase the contribution of the OSH function.
- Consider some additional considerations to assist in explaining how an enhanced OSH strategy can benefit the organisation.
- Explain some organisational processes that can be used by OSH professionals to increase their involvement and influence throughout the organisation.

7.79   What are the key objectives of a strategy? We suggest they are:

- To influence the inclusion of 'risk management' on the broader business agenda, particularly at board level, and influence corporate governance and internal control processes.
- To assist management in developing the organisation to ensure that it effectively balances the management of its opportunities, whilst minimising its risks, thereby supporting the achievement of its strategic objectives.
- To develop the role of OSH professionals, and the acceptance of the role change, to one of managing risk within the total range of OSH activities, and to increase the added value of the OSH function across its range of activities.
- To increase the influence on the business by changing the OSH role from legal compliance to business partnership.

RISK ASSESSMENT AND CLASSIFICATION

7.80   Where to start? This phrase springs to mind – 'Get a quick win', eg, tackle a situation:

- Where the risk is self-evident.
- Where the assessment can be undertaken using existing evidence, and the priority is likely to be high.
- Where risk treatments are available to the OSH function.
- Where risk controls can be implemented.
- Where the resolution of the issue can have a quick and effective impact, especially if a reduction in costs is a real possibility.

7.81   However, whilst we appreciate that it is very tempting to 'rush off' to tackle a particular risk, we strongly recommend that a longer-term perspective is taken and that the risk assessment process is tackled in a systematic manner. In that way, those who OSH professionals seek to influence, and those who will need persuading that OSH is part of risk management, will see that there is some substance to the approach.

7.82   Tackling some risks is going to take time and the creation of a carefully structured strategy and plan. There are no 'quick-fixes' that are going to resolve certain risks overnight, especially those that relate to organisational factors and OSH risks.

7.83   We recommend that a structured risk-assessment process be used. In addition to considering the typical 'probability', 'impact', and potentially the 'number of people affected' variables, we recommend that consideration is also given to the cost-reduction opportunities for each risk, and the cost of the intervention. This will create a 'cost-benefit' analysis approach that will show that there is a business and commercial focus to the analysis, which will appeal to many within the organisation. An additional 'selling point' that we have found very successful is to translate the cost of inaction into the cost of providing a service or the profit on a particular product. We find that this concentrates the mind of business managers very well, if they understand that the cost of inaction is actually higher than the cost of the intervention, especially if the cost, eg the uninsurable costs arising from accidents, is more than the profit being achieved on the product/service that the manager is responsible for. It becomes even more significant where the uninsurable cost is charged to their cost centre, and they do not have a budget for those costs.

Classification process

7.84   When the risk-assessment process is completed, and the OSH function is aware of the level of risks, there will be a need to prioritise the risks for implementation and building into the strategy. Whilst all risk-assessment processes do provide a 'ranking order', often the risks to be tackled first are decided by other, broader considerations, including the availability of resources – financial, people, time – to implement the risk treatment(s) and risk-control methods identified. Consequently, we recommend that a simple process be used to consider each item listed in the Risk Register and to establish an overall classification. This process should be undertaken by asking these questions:

1   Is it within the OSH professional's authority and responsibility to implement the risk treatments option(s) listed?
2   Does the OSH professional have the current resources – money, people and skills – to implement the option(s)?

7.85   If the answers are 'yes', then consider those as *priority 'A' actions*. The implementation of these items, and the communication of the action and the benefits in managing the risks will start the process of the rest of the organisation realising that the OSH function is also in the business of 'risk management' and is equally as focused on business and commerce as other parts of the organisation.

7.86 If the answer is 'no' to either of the above questions, then there is a need to ask some further questions:

1   Could the risk treatment option(s) listed be implemented by agreeing the option(s) with the OSH professionals' immediate superior – either line or functional – or 'local' management team?
2   Can the immediate superior – either line or functional – or 'local' management team agree to provide the resources – money, people and skills – to enable the option(s) to be implemented?

7.87 If the answers are 'yes', then consider those as your *priority 'B' actions*.

7.88 If the answer is 'no' to either of the above questions, then you need to ask some further questions:

1   Could the risk treatments option(s) listed be implemented by agreeing the option(s) with the ultimate authority within the organisation, ie chairman, chief executive, board of directors?
2   Will the ultimate authority within your organisation, ie chairman, chief executive, board of directors, agree to provide the resources – money, people and skills – to enable the option(s) to be implemented?

7.89 If the answers are 'yes', then consider those as your *priority 'C' actions*.

7.90 If the answer is 'no' to either of the above questions, then those actions are your *priority 'D' actions*, but do not delete them from your Risk Register. Remember there may be an opportunity to input your ideas at a later date, especially if you have successfully implemented your 'A', 'B' and 'C' actions.

7.91 However, this simple approach should not restrict consideration of any of the risk treatments as a higher priority. Senior management may decide that certain risks should be tackled first, even if they are more of a challenge than, say, priority 'A' risks, and more resource intensive, simply because they are more strategic and can deliver a better added value to the achievement of organisational objectives.

Additional considerations

7.92 Referring again to the classification system described above, there will exist a classification for each risk detailed in the Risk Register and a justification for each change – or, as a minimum, for the priority 'A' items.

7.93 We suggest that a valuable input to the development of the strategy is to ask key managers and employees within the organisation what aspects of OSH risks they think the organisation should focus on. The answers have to be treated with caution because some managers will respond in a manner that indicates that they want to opt for the stance that the organisation does not need any more management of its OSH risks, as that would encroach on their existing risky method of working.

7.94 Asking your 'client groups' – managers and employees – will, in theory, enable changes to be implemented with limited resistance, but may restrict changes to those that the organisation is prepared to accept. However, it is at least a move in the right direction.

7.95 So what is your strategy for moving forward from this point? We suggest that, in general terms, you have three options:

1   Proceed with the changes on a piecemeal basis – priority 'A', then priority 'B' and so on.
2   Identify the changes that will be required to implement priority 'C' items and develop a plan to change the thinking within the organisation to implement those changes.
3   Combine the two options; by starting on priority 'A' items, but at the same time start a process of discussion with other members of the team about the priority 'C' items.

7.96 Clearly your decision on an option will be conditioned by organisational, operational, professional and personal factors that we have discussed previously. In general terms, organisational factors are more likely to be priority 'C' items, operational items priority 'B' items, and professional and personal factors priority 'A' items. In addition, it is generally easier to determine and explain your strategy if it is structured in a systematic manner, and/or is explained as a model. We have set out some ideas below.

7.97 One way to look at your strategic contribution is to view it as a model (see **Figure 7.2**), with the organisation being shown as a system for producing a product or service to create a profit for shareholders.

FIGURE 7.2

7.98   The model shows that if the organisation ignores the contribution that can be made by the management of OSH risks, then directly or indirectly the combination of resources + process can affect the output of products and services that create the profit for the organisation.

7.99   Another way to look at it is to consider each major function of the organisation as a system. However, as we have discussed earlier, if the organisation and each separate activity does not adopt an integrated process approach, then each activity will operate within its own silo. What do we mean by 'silo'? The term means that each area will only view its contribution to the organisation within its own boundaries, which are generally seen as a vertical process. Consequentially, each function tends to operate in isolation from any other and will not appreciate or concern itself with the concept of 'input' and 'output' risks. Typically, such organisations are very bureaucratic and operate along rigid vertical reporting lines. The tendency is for contact to be made between the functions via the vertical 'chain of command' rather than a more flexible arrangement, where people are encouraged to work with anyone within the organisation who can support their objectives – sometimes called a 'matrix approach'.

7.100   Viewed another way, this can produce an organisation that looks like this.

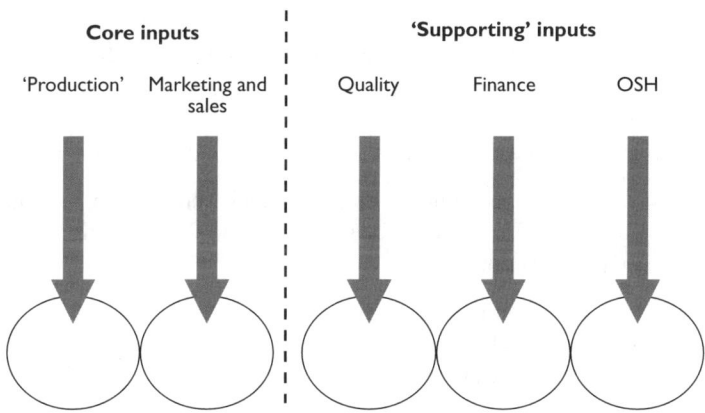

FIGURE 7.3

7.101   However, the consequences of a rigid, silo-focused organisation are significant. Each function concentrates on its own activities and does not consider the benefits of a more co-operative organisational system approach. We do not have the space in this book to debate the benefits and consequences of each type of organisation structure, but in our experience the following approach (see **Figure 7.4**) is significantly more beneficial for all stakeholders. This creates an organisation where each activity overlaps or is co-ordinated

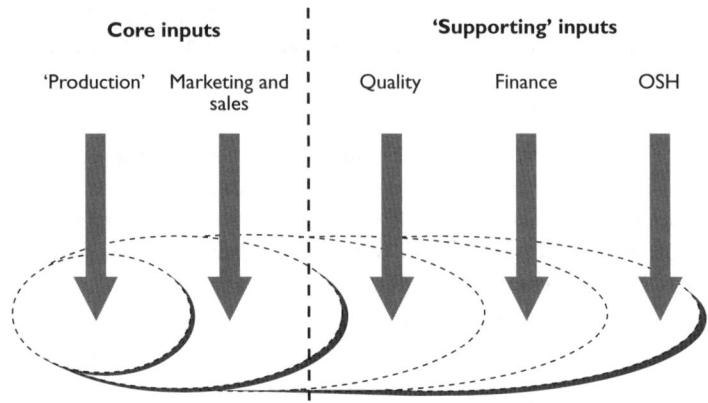

FIGURE 7.4

with each other. In this example, 'input' and 'output' risks are identified both within each area but also between areas.

Other strategic development opportunities

7.102 We have identified several organisation-wide internal control systems that provide an excellent opportunity for OSH professionals to increase their involvement, contribution and influence. By deciding to involve themselves in these systems, OSH professionals can build these activities into the OSH strategy and demonstrate that the strategy has considered all the factors we discussed earlier in this chapter.

7.103 We have provided outline details of each and explain how they can become part of the OSH strategy for increasing the OSH functions contribution to the management of risk:

● Corporate governance (CG).
● Corporate social responsibility (CSR).
● Operating and financial review (OFR).

Corporate governance (CG)

7.104 We first discussed CG in **Chapter 1** and then in **Chapter 2**, In summary, CG relates to the senior-level internal control processes of an organisation, and in particular the arrangements for managing risks. In the case of the private sector this includes:

● Guidance for the appointment of the Chairman, directors and non-executive directors, their roles and responsibilities.
● The creation of a system for internal control.

- The auditing of the board's performance.
- The principal duties of the remuneration and nominations committees.

7.105 The Combined Code on Corporate Governance, issued in July 2003, includes matters taken from the Turnbull Report mentioned in **Chapter 1**. The Guidance in the Turnbull Report makes reference to the need for a sound system of internal control in the particular circumstances of the organisation (company).

7.106 OSH professionals have an opportunity to conduct their activities in accordance with the principles of internal control and to make an effective contribution to the organisation's overall internal control systems.

7.107 The Guidance states:

'All employees have some responsibility for internal control as part of their accountability for achieving objectives. They, collectively, should have the necessary knowledge, skills, information and authority to establish, operate and monitor the system of internal control.'

7.108 OSH professionals can make a valuable contribution to CG by seeking to influence the organisation's strategy through contact and discussion with the directors and non-executive directors. The information that OSH professionals have included in the Risk Register, and the output from the other processes, will be of particular interest to those internal control systems that are a major concern of the organisation. In addition, the list of priority 'C' actions (these, if you recall, are likely to be the organisational-level items) can help the organisation to understand and focus on the organisational-level risk areas that are likely to be having a significant impact on the ability of the organisation to achieve its strategy, goals and objectives.

7.109 The OSH strategy should include an analysis of the opportunities for increasing influence in corporate governance systems, either directly or indirectly, and use this analysis to produce an action plan.

Corporate social responsibility (CSR)

7.110 CSR is essentially about how an organisation takes account of its economic, social and environmental impacts in the way it operates, maximising the benefits and minimising the downsides.

7.111 In May 2003, the UK Department of Trade and Industry and the Forum for the Future hosted a workshop for business, non-government organisations, academics and government to debate the links between competitiveness, productivity and the increasingly important role of intangible assets, as well as sustainability and CSR. The broad conclusion, with some qualification, was that sustainability makes a positive contribution to business success.

The key is to look at CSR as an investment in a strategic asset or distinctive capability, rather than an expense. The debate highlighted the importance of taking a balanced approach to assessing performance, rather than concentrating solely on one aspect, such as shareholder value.

7.112   Work is continuing to identify the links between CSR and organisational effectiveness and success, and there are some examples, eg British Telecommunications, of companies that have carried out a lot of work in this area, with particular reference to reputational risks.

7.113   There are clear links to the management of OSH risks, the activities of the OSH function and the development of a strategy to increase the influence of OSH. For example, how would the organisational agenda for CSR be communicated throughout the organisation, so that the CSR policies are an integral part of the way that the organisation is managed? How would individual practices and behaviours be brought together to ensure that the CSR policy is consistently maintained? How could performance-management systems discuss and recognise employee CSR initiatives and success? How can incentives be used to focus managers and employees on the CSR initiatives? The relevance to the OSH function is that the same questions are relevant to the management of OSH risks. Consequently, if the OSH professional gets involved in the organisational process to answer these questions for CSR purposes, there will be a spin-off for OSH risk management.

7.114   OSH professionals should seek to influence the CSR debate within an organisation and identify ways of supporting the implementation of CSR policies. The OSH strategy should, therefore, include an analysis of the opportunities for increasing influence in CSR policies and initiatives, either directly or indirectly, and then using this analysis to produce an action plan.

Operating and financial review (OFR)

7.115   In May 2004, the government published a consultation document on 'Draft regulations on the Operating and Financial Review and Directors Report'. This was in response to the recommendations of the independent Company Law Review and the Accounts Modernisation Directive from the European Union, and other inputs.

7.116   Paragraph 2.5 of the Draft Regulations in part states:

'Directors deciding in good faith what would be most likely to promote the success of the company, taking account of a wide range of factors, within and outside the company, which are relevant to achieving its objectives and to an assessment of the business. These factors may well include the company's impact on the environment and on the wider community, and its relationships with employees, customers and suppliers.'

7.117   We believe that there is a clear implication that the management of OSH risks should be a key component of this review and report, and OSH professionals should seek that the management of OSH risks are considered and comment included in the review and report.

7.118   When finalised, the requirements will apply to financial years starting on or after 1 January 2005 (since amended to 1 April 2005) for large and medium-sized companies.

7.119   Many organisations are currently reviewing their response to the OFR and, in general, see the OFR as a progressive move, but it is unclear at this stage whether it will actually create a willingness and appetite to take the management of OSH risks more seriously. It will largely depend on whether the director who signs off the OFR actually thinks that OSH issues are a real risk to the business and shareholder value. OSH professionals can assist and contribute by asking the questions set out below. OSH processes need to be incorporated efficiently and effectively into the organisation's risk management process in order to answer many of the questions being raised.

7.120   OSH professionals can assist and contribute to the thinking and the preparation of the review and the report's contents by posing the following questions.

- Does the organisation currently undertake such a review and publish such a report?
- Does it include aspects related to the management of OSH risks?
- Are OSH risks a significant aspect of your organisation?
- Has the organisation started to review its current practice to take account of the likely changes?
- Have OSH professionals considered the matters relating to the management of OSH risks that could be appropriate for inclusion in the review and report?
- Have OSH professionals been involved in the broader discussions?
- Have OSH professionals sought out and created new internal 'partnerships' to ensure that their contribution is timely and effective?

7.121   The answers to these questions are particularly important, as the consultation document does not indicate that any particular approach will be required.

7.122   However, paragraph 3.33 of the draft regulations states:

'A poor record on environmental or OSH matters, for example, could adversely affect a company's standing and business prospects. For regulated sectors, non-compliance could lead to the loss of licence to operate, and in some cases, imprisonment for directors. The OFR requires such matters to be covered, both where they constitute a significant external **risk** to the company, and where the company's

impact on others through its activities, products or services, affects its performance. For example, information, which could be necessary for shareholders to make an assessment of the business includes:

● An explanation of risk management approaches employed by a company that stores, transports or uses significant volumes of hazardous or toxic substances that risk damaging the health of workers or others, or polluting the environment or;'

7.123 Paragraph 3.34 goes on to say:

'The financial loss to the company from poorly managing these issues could be direct (eg through fines, increased material costs, increased labour costs), indirect (through loss of reputation, supply failure, production interruption, property damage, customer loyalty) or from the costs associated with missed opportunities. Failure to anticipate or influence the changing regulatory landscape could also affect long-term performance.'

7.124 The above statements provide a significant opportunity for OSH professionals to get on the 'risk and business agenda'. Consequently, OSH professionals should use the outputs from their risk identification process, and the information from the other strategic development opportunities outlined above, to discuss and influence directors and senior management about the content of the OFR, relating to the management of OSH risks.

IMPLEMENTATION CHALLENGES

7.125 At this point a great deal of information will be available:

● OSH Risk Register.
● Risk assessments.
● Classifications.
● Selected risk treatments and controls.

7.126 The information should be discussed and agreed with the OSH team. The risk treatment and risk controls should be appropriate and proportionate. However, before any decisions are made about 'risk treatment' options, or decisions made about 'risk control' measures, you will need to identify the obstacles to implementation.

7.127 OSH professionals should seek out and create business partnerships with those parts of the business that have influence through other mechanisms, eg financial control, legal, and company secretariat. In that way, it will increase its influence via direct and indirect routes. In our experience, if OSH professionals can establish a partnership with the finance function, then there is very little that anyone in the organisation can do without needing to speak to one or other of the partnership! If they create an effective working partnership, then there is an excellent chance that the risk agenda will be welcomed and more readily accepted.

7.128 By using the classification system described above, the implementation challenges will have been anticipated.

7.129 What is now needed is to develop a formal justification for the implementation of each risk-treatment and risk-control option. We suggest that this is even undertaken for the priority 'A' items, even though it has been decided that the change can be implemented without reference to others or the need for additional resources. This is for several reasons:

1   It will provide practice in thinking in 'risk' terms and thinking of the improvement in managing OSH risks.
2   It will provide practice in determining the benefits to business and commercial processes, and what contribution it will make to improving the management of the organisation's risks.
3   It will provide readily available answers, should they be required, to enable a start to be made in the process of increasing influence.
4   It will justify the change should the organisation decide that resources need to be reduced, costs to be cut, etc.

7.130 Creating an OSH Risk Register, using a classification system to determine priorities, and developing a rationale for each risk treatment and control option will not necessarily result in unqualified success on every occasion. There are some implementation challenges that will need to be considered and that can be the most significant challenges to overcome. This is especially relevant if the decision to embark on an OSH risk management programme is the first major excursion by the organisation into the world of risk management, other than the traditional insurance function.

7.131 In our experience, the 'first' major excursion into a new area of internal management control, be it quality, IT, environment, human resources, customer service or OSH is the most challenging because a control mentality is often not part of the organisation's management style. Most organisations that have not faced a new, potentially all-embracing management control process will find it difficult to accept the actual change and the need to implement controls which, at first, seem to be bureaucratic. The 'second' management control process that is introduced will proceed much easier because an element of the 'control mentality' from the first process has become ingrained in the organisation; thereby reducing the level of resistance to change that happened on the first occasion.

7.132 We recommend that the following aspects be considered during the implementation of any new risk-management process, and especially during any move to increase the influence and involvement of the OSH function:

1   What is the extent of the change and what impacts will the change have outside of the OSH function?

2   Is there available an effective internal communication process that can be utilised to communicate the change and the supporting information?

3   While senior and middle management may support the concept of risk management, it requires discipline, time and effort to implement a regular, systematic process. Managers need to truly feel that risk management work is of value and understand that it is not an add-on, but a better way of doing business.

4   The start-up phase of a risk-management process can require some time and effort to overcome the inertia of current practices. However, as risk management becomes integrated with management systems and business processes in general, people will realise that additional resources may not be required.

5   Senior management should be required to report progress on the implementation of risk management processes, so that the momentum is maintained and managers are held to account for their agreed actions.

6   The implementation of risk management processes will impact on both priority setting and longer-term plans to achieve organisational objectives

FINAL STAGE

7.133   None of the above methods provide all the information needed to create and implement a strategy for OSH professionals to increase their contribution to the management of risk within the organisation, and especially OSH risks. However, each can provide valuable assistance at a different level and at a different stage in the process of building the strategy. Therefore, OSH professionals should use a combination of approaches or tools that best fit the organisation, closely reflect the current status of the OSH function and provide the insights and outputs that are desired.

7.134   At this stage, the following information is potentially available:

• An OSH Risk Register.
• The output from one or more of these approaches or tools:
    – RiskFrisk® – OSH Risk Management.
    – Corporate governance.
    – Corporate social responsibility.
    – Operating and financial review.
• An outline action plan with priorities.

7.135   A combination of the outputs from the above approaches and tools will provide OSH professionals with sufficient information and supporting evidence to create a draft strategy.

7.136 The draft strategy should then be discussed with the OSH team, the 'business partner(s)' or line/functional manager to agree an actual action plan, with priorities, timescales and measurement criteria. The plan should include adequate monitoring processes and regular reviews of the progress of implementation.

REFERENCES

Code of Professional Conduct, The Chartered Institution of Occupational Safety and Health (IOSH) (http://www.iosh.co.uk)

Combined Code on Corporate Governance – UK Financial Services Authority, 2003. The Financial Reporting Councils' Combined Code – Listing Rules for UK Stock-Exchange listed companies (http://www.frc.org.uk)

Draft Regulations on the Operating and Financial Review and Directors' Report – a consultation document – UK Department of Trade and Industry, May 2004 (http://www.dti.gov.uk/cld/condocs.htm)

Forum for the Future (http://www.forumforthefuture.org.uk)

Turnbull Report, *Internal Control: Guidance for Directors on The Combined Code*, The Institute of Chartered Accountants in England and Wales, September 1999 (http://www.icaew.co.uk/internalcontrol/)

UK Department of Trade and Industry and the Forum for the Future, workshop for business, non-government organisations, May 2003 (http://www.societyandbusiness.gov.uk)

CHAPTER 8

Pulling it together

8.1 This chapter will pull together the key points from earlier chapters and summarise the way forward for OSH professionals to identify and assess OSH risks.

8.2 The theme of this book has been that where OSH professionals operate at a strategic level within the organisation's strategic processes and business and operational processes, they are ideally placed to make a significant and effective contribution to organisation-wide risk management. Risk management succeeds or fails based on altering employee's perceptions, attitudes, behaviour and performance with regard to risk. Success will depend on effective training, performance management, reward and sanction systems and developing work practices and procedures that limit human error, increase job satisfaction and reduce stress. The above mechanisms are cross-functional, and interrelated, hence the need for the OSH professional's involvement at the strategic level, with a cross-functional remit.

8.3 In **Chapter 2** we explained that the management of risk is a vital and key part of managing any organisation. We used a brief description of some important cases – *Herald of Free Enterprise*, Piper Alpha, etc, to show how each had a significant OSH element that was not taken into account during more normal 'insurance/financial' focused business risk management processes to identify and control business and operational risks. We also pointed out that if a business risk management programme had been in place and meaningful risk decisions had been taken further up the management tree with all relevant risk assessments in place, rather than merely making decisions on an arbitrary cost basis, then each case could have been foreseen and preventive actions taken.

8.4 We explained that this so-called high-level 'cost only' decision-making is symptomatic of many board decisions taken without much thought for the risk side of the equation. However, a business risk management (BRM) programme should highlight the importance of risk assessments to the board/senior management within organisations and ensures that both cost and risk are taken into account when management decisions are taken and implemented.

8.5   We showed how the consideration of non-financial/operational risks at board level has been elevated in importance over the last few years, with Turnbull being seen as something of a watershed.

8.6   We discussed how a BRM programme helps to elevate the profile of OSH within an organisation's overall corporate governance (CG) and corporate social responsibility (CSR) approach, and stressed the need for the 'safety net' to be extended to include all potential organisational stakeholders.

8.7   We examined the role that OSH professionals can play in BRM and also highlighted how a proactive BRM programme can greatly assist organisations in achieving their CG and CSR goals and objectives. This is especially important as the UK Health and Safety Commission (HSC) and UK Health and Safety Executive (HSE) are taking an increasing interest in how Stock-Exchange-listed companies are reporting on CG/CSR/OSH issues in their Annual Reports.

8.8   We showed how effective corporate governance processes within an organisation must include OSH and how business decisions must consider all risks and consequences of a business strategy. We explained that it is traditionally difficult for OSH professionals to participate in CSR and CG processes, as they often only operate at the 'operational' level within organisations and, additionally, approach their roles in a risk-averse, non-business added value manner. We demonstrated how OSH professionals can seek to increase their influence 'up the management chain' so they are asked to contribute at the 'tactical' and 'strategy' levels, where business risk management is typically on the agenda. In this way, OSH professionals can help the organisation to manage its opportunities in a more complete manner, whilst minimising the risks.

8.9   It is vital in the context of CG, CSR and BRM that OSH professionals learn the language of the boardroom so they are invited to participate in the CG/CSR/BRM processes.

8.10   As organisations get to grips with the increasing need to encompass all three relatively new business requirements, the OSH professional will have an increasingly important role to play at board level in order to ensure that organisations fully adopt the BRM/CSR/CG principles and processes. In so doing, organisations will find themselves moving towards the ultimate goal of continual improvement in all their business performance indicators, including continual improvements in their OSH management systems.

8.11   **Chapter 3** shows that OSH risk management systems are vital in terms of ensuring compliance with the requirements of CG, BRM and CSR principles, concepts, processes and best practices. As with any management system, they need to be fully integrated into the normal business and operational processes of the organisation – built in, not bolted on!

8.12 We included an overview of existing management systems and explained the parallels between quality management systems and OSH management systems, and the changes to BS 8800. We described the benefits to organisations of integrating their management systems, rather than having separate systems and discussed the auditing/verification process.

8.13 We explained that a systems approach is by far a better, proactive way of managing all business risks, including OSH business risks, and we provided an explanation of 'management systems' and the benefits in managing OSH risks by using a systematic approach that is linked to the needs of the organisation and its business and operational processes.

8.14 We explained that it is imperative that the OSH professional understands and talks the language of the boardroom so that an OSH risk management system is accepted as part of normal business and operational processes. This ensures that OSH considerations are taken into account on a cost versus risk basis, so that the business case for OSH risk management is made using both sides of the cost versus risk equation. On one side of the equation are the costs – which may or may not be losses – and on the other side are the profits (including cost savings) that emanate from effective, proactive business risk management – ie commensurate risk-control systems.

8.15 Within the OSH arena, there are legislative requirements to take into account. These force organisations to demonstrate – beyond reasonable doubt (statute law) – that they have taken into account both sides of the equation in terms of 'so far as is reasonably practicable (SFARP)'. As stated in **Chapter 2**, only when the cost is grossly disproportionate to the risk is it not reasonably practicable to ensure that suitable and sufficient risk control systems – as part of an overall OSH risk management system – are in place and are operational.

8.16 It is therefore important for OSH professionals to be able to quantify both the cost of loss (ie the risk actually resulting in a loss) and the cost of risk prevention (ie the control measures) in economic terms, rather than just stating 'we have to comply with what the law says' which, inevitably, is a poor motivator, especially at board level. This is largely because most boards are remote from the action at the operational levels of their organisations and, although they are considered to be the controlling mind of the organisation, they are – especially in larger organisations – also remote from the risk of individual prosecution under OSH legislation, as cases – including some discussed in **Chapter 2** – have illustrated. Hence, there is a push from pressure groups and many sectors of the general public to implement some form of corporate manslaughter/reckless killing legislation that enables boards to be directly responsible for violations of OSH regulations that lead to direct prosecution for corporate manslaughter. The 'controlling mind' of smaller organisations is more easily identified, hence successful prosecutions of directors of smaller companies. However, directors of larger companies regard themselves as immune from prosecution, until the law changes. When that happens we anticipate that boardroom attitudes may rapidly change.

8.17   Having said that, the Health and Safety at Work etc Act 1974, s 37 – highlighted in **Chapter 2** – is still in force but has only been used to prosecute directors and senior managers in small organisations where the linkage evidence is almost self-evident.

8.18   **Chapter 4** made reference to the Health and Safety Executive's publication *Successful Health and Safety Management* – HSG 65 (1st Edition, 1991) – which gave rise to the mnemonic 'POPIMAR' which we used as the framework for this chapter.

8.19   In **Chapter 5** we described the process for implementing OSH business-related processes. **Chapter 4** provided a legal and overall backdrop to the subject of OSH business integration set out in **Chapter 5**.

8.20   POPIMAR was used to describe the key stages in the development and implementation of an effective and efficient OSH policy that should be the cornerstone of any OSH management system.

8.21   We showed that all stages are integrated to create a closed-loop process that feeds back requirements/suggestions for continual improvement to the policy formation and review stages. If the board and senior management are not provided with regular feedback, they will wrongly assume that everything is in order and going to plan. This is unfortunately the case with reactive OSH risk management that all too often results in a major accident, before anything constructive takes place.

8.22   The POPIMAR approach to OSH policy and OSH risk management systems clearly demonstrates that the organisation has taken on board CG, CSR and BRM concepts and practices in their overall management of risk.

8.23   We concluded by explaining that the best OSH policy is one that is implemented in an integrated manner, that fits the requirements of the organisation, works in practice and one to which the organisation is visibly and demonstrably committed at all times, in all operations, in all locations, and at all levels.

8.24   **Chapter 5** showed how the implementation of an OSH risk management system could be integrated into the organisation's normal business and operational processes.

8.25   Earlier chapters had discussed key building blocks in the process to make OSH part of business risk management. In **Chapter 5** we concentrated on the OSH business processes that we have found can make a big difference to the successful integration of OSH risk management within an organisation's 'business and commercial environment' and business and operational processes.

8.26 We explained how OSH professionals need to view their organisation, or client organisations, as a complete system so that business processes complement one another and are designed to ensure an integrated, consistent and non-duplicating approach. We used our experience to show that this approach is appreciated and welcomed by organisations who are generally looking for flexibility, added value and not uncoordinated 'red-tape'. They respond much better to the use of business and commercial focused interventions, and can see the added value of good/best practice if it is explained in business terms.

8.27 It was explained that the implementation of OSH business processes was designed to ensure that OSH risk management requirements are integrated with normal organisational processes. If OSH processes are established as part of the way that the organisation runs its activities, then OSH will be managed as a normal part of management and employee activities. Management and employees must be heavily involved in the design, implementation and ongoing monitoring of these processes. The organisation's OSH professional should not be required to be the management systems policeman, but should focus on advising management and employees, in order to minimise the need for constant 'firefighting', thereby resulting in proactive resource allocation to ensure continual improvement.

8.28 In **Chapter 6** we saw how organisational psychology provides a valuable perspective in managing organisational risks. We also saw how this approach to risk management has evolved to provide OSH professionals with a key perspective in understanding the causes and motivations associated with risk-related behaviour in the workplace.

8.29 We considered two perspectives, which are brought together to demonstrate an integrated approach to risk management. First, organisational factors – the structures and processes that determine the culture of an organisation and include factors such as responsibility frameworks, communication frameworks and job/role design. Second, human factors – the perception of risk, the nature of human error and what motivates individuals in the workplace to take risks.

8.30 OSH professionals operating at a strategic level are best placed to manage the risks associated with organisational and human factors. They alone can take a cross-functional view of organisations. The focus of OSH professionals on people management through such activities as training, performance management and reward mechanisms provides them with the opportunity to become an integral and vital part of the risk management process.

8.31   The input of OSH professionals to the risk management process is about implementing the procedures and practices that lead to changes in behaviour, attitudes and values. A systematic focus on limiting human error at every stage and process within an organisation acts as a mechanism for organisational development as the potential for human error is largely synonymous with inefficient organisational systems.

8.32   Learning and development are key components in establishing a positive risk culture. Ineffective learning (or 'training') is a fundamental cause of loss and accidents in the workplace. OSH professionals must seek to implement a risk-management system. As with any organisational change, this will require training to provide management and employees with the necessary skills, knowledge, and attitudes to achieve the organisation's new strategic goals.

8.33   It is difficult, if not impossible, to eradicate human error from organisations. However, OSH professionals can do a significant amount to reduce the risk of human error, by influencing and supporting organisational systems that seek to limit risk-related behaviour.

8.34   The recurring theme throughout the book is the strong emphasis on OSH professionals raising their profile and increasing their contributions within the organisation. The various chapters have examined the risk of the OSH professional being unable to achieve this. **Chapter 7** focuses on providing guidance as to how OSH professionals can increase their influence, and contribute significantly towards the performance and development of the organisation. The chapter shows how to develop a strategy for increasing the OSH contribution to the management of risk within the organisation.

8.35   Details are included of a structured process for identifying and classifying OSH risks, and how the results of the process can be built into a Risk Register. The concept of the Risk Register is introduced so that all identified OSH risks are recorded and can then be classified to ensure that all potential risks are dealt with in an appropriate order of priority.

8.36   Four factors are used to describe the process:

1   OSH organisational and strategic risks (organisational context within which the organisation is operating, eg identifying the organisation's activities, their location, the ownership status). These aspects vary considerably from organisation to organisation and affect the organisational impact on people risks. OSH professionals need to identify the factors that are relevant to their organisation, evaluate the impact, and record the risks in the Risk Register.

2   Policies and systems for general risk management. OSH professionals must develop a consistent approach with existing 'risk management' functions to avoid conflict, misunderstanding or confusion.

3   OSH strategic, tactical, operational, professional and personal risks.

An OSH Initial Status Review (ISR) may need to be carried out in order to identify such risks, plus develop a continuing professional development process for professional and personal risks.

4   Risks generated by the organisation's business and operational processes. All the organisation's business and operational risk areas which impact on the management of OSH risks must also be incorporated.

8.37 Details of a unique risk-profiling tool that has been developed especially for OSH professionals are included. The tool is called 'RiskFrisk® – OSH Risk Management' and it is aligned closely with the above four factors. The unique tool focuses on the way in which an organisation's risks and the interrelationship between them impacts on the management of its OSH risks.

8.38 In addition, an explanation is given of a process for developing a strategy to enable OSH professionals to use organisational processes to increase their influence. There is a strong focus on processes to identify risks so that OSH professionals have the opportunity to influence the organisation at all levels.

8.39 It was explained that none of the methods described would provide all the information needed to create and implement a strategy for OSH professionals to increase their contribution to the management of risk within the organisation, and especially OSH risks. However, each could provide valuable assistance at a different level and at a different stage in the process of building the strategy. Therefore, OSH professionals should use a combination of the approaches or tools that are a best fit for the organisation, closely reflect the current status of the OSH function and provide the desired insights and outputs.

8.40 However, a combination of the outputs from the approaches and tools described will provide OSH professionals with sufficient information and supporting evidence to create a draft strategy. We recommended that the draft strategy should then be discussed with the OSH team, the 'business partner(s)' or line/functional manager to agree an actual action plan, with priorities, timescales and measurement criteria. The plan should include adequate monitoring processes and regular reviews of the progress of implementation.

CONCLUSION

8.41 Through the use of this book we believe that OSH professionals will be able to carry out a complete review of the OSH function and its key areas of impact on the organisation. OSH professionals will then be able to undertake risk identification and effectively record the risk and potential risk treatments to minimise risks and maximise opportunities using a structured approach. A structured risk-identification and risk-treatment process will provide a valuable opportunity for OSH professionals to demonstrate their risk credentials.

8.42   Through such an approach the OSH professional will begin to understand the need to develop an OSH risk management strategy linked to the organisation's overall risk management strategy. Such strategic level involvement allows the OSH professional to demonstrate the value of its increased contribution to the organisation. The organisation should then take advantage of this enhanced contribution to ensure that OSH risks are managed effectively and that the contribution by the OSH function is maximised to the benefit of the organisation and all its stakeholders.

Index